THE PHYSICIST'S
CONCEPTION OF NATURE

The Physicist's Conception of Nature

WERNER HEISENBERG

Translated from the German
by
ARNOLD J. POMERANS

Harcourt, Brace and Company

New York

©1955 by Rowohlt Taschenbuch Verlag, GmbH.
©1958 by Hutchinson & Co. (Publishers) Ltd.

*All rights reserved. No part of this book may be
reproduced in any form or by any mechanical means,
including mimeograph and tape recorder, without
permission in writing from the publisher.*

first American edition

Published in Germany under the title
Das Naturbild der heutigen Physik

Library of Congress Catalog Card Number: 58-10896

PRINTED IN THE UNITED STATES OF AMERICA

CONTENTS

1 THE IDEA OF NATURE IN CONTEMPORARY PHYSICS *Page* 7
    The Problem of Nature
    Technology
    Science as a Part of the Interplay between Man and Nature

2 ATOMIC PHYSICS AND CAUSAL LAW 32
    The Concept of Causality
    Statistical Laws
    The Statistical Character of Quantum Theory
    The History of More Recent Atomic Physics
    Relativity Theory and the Dissolution of Determinism

3 CLASSICAL EDUCATION, SCIENCE AND THE WEST 51
    The Traditional Reasons for the Defence of a Classical Education
    The Mathematical Description of Nature
    Atoms and Classical Education
    Science and Classical Education
    Faith in our Task

CONTENTS

Historical Sources

4 THE BEGINNINGS OF MODERN SCIENCE 71
  Johannes Kepler
  Galileo Galilei
  Isaac Newton

5 THE ORIGINS OF THE MECHANISTIC AND MATERIALIST WORLD-VIEW 121
  The Application of Newton's Methods:
  Christian Huygens—Gassendi—
  Boyle—Descartes
  D'Alembert
  De la Mettrie
  Wilhelm Ostwald

6 THE CRISIS OF THE MECHANISTIC-MATERIALIST CONCEPTION 152
  Heinrich Hertz
  Louis de Broglie

Summary 180
About the Author 182
Bibliography 185
Name Index 191

I

THE IDEA OF NATURE IN CONTEMPORARY PHYSICS

IT HAS been said that the modern attitude to nature is so different from that of previous times that all our relationships to her, even those of the artist, must now be based on a new set of premises. Whereas in previous centuries man's attitudes were expressed in a rounded natural philosophy, today they are largely determined by modern science and technology. Thus it is important—and not for the scientific investigator alone—to examine the idea of nature in contemporary science and particularly in contemporary physics. Now, while there is little reason for thinking that modern science has had a direct influence on, say, the artist's discourse with nature, the basic changes in modern science must yet be considered as expressions of changes in our very existence and thus as affecting every realm of life. If this be the case, even those who try to fathom the essence of nature creatively or philosophically must take notice of the changes in the scientist's idea of nature that have taken place during the last few decades.

The Problem of Nature

Changes in the Investigator's Attitude to Nature

Let us first look back at the historical roots of modern science. When it was founded by Kepler, Galileo and Newton in the seventeenth century, there still prevailed the idea of nature of the Middle Ages. Nature was seen as God's creation. Nature was God's Work and it would have been thought senseless to ask questions about the material world without reference to God. As a document of the times, I should like to quote Kepler's concluding remarks in his *Cosmic Harmony*: 'Thanks be to Thee, Oh Lord our Creator, who hast granted me visions of beauty in Thy creation, and with the work of these Thy hands I give praise. Lo, I have completed the work to which I have been called; I have exploited the gifts Thou hast bestowed upon me; I have proclaimed the splendour of Thy work unto those who will read these proofs, in as much as I, in the limitations of my mind, have been able to grasp them.'

Yet, within the course of a few decades, man's attitude to nature was to change radically. As scientists delved more deeply into the details of natural processes they realized, as in fact Galileo had been the first to do, that individual natural processes can be isolated from their context in order to be described and explained mathematically. At the same time it became clear how immense was the task confronting this new science. Thus, Newton no longer looked upon the world as a whole that could only be understood as God's work, and his attitude to nature is best summed up by his well-known statement: 'I do not know what I may appear to the world, but to myself I seem to have been only like a boy playing on the seashore, and diverting myself in now and then finding a smoother pebble or a prettier shell than ordinary, while the great ocean of truth lay all undiscovered before me.'

This change in the scientists' attitude to nature is perhaps best understood if we consider that, to Christian thought of the time, God seemed to be in a Heaven so high above the earth, that it became significant to look at the earth without reference to God. In this respect we may even be justified in speaking of modern science—as Kamlah appears to do—as a specifically Christian form of Godlessness, and so explain why there has been no corresponding development in other cultures. It is hardly accidental that in the creative arts of that time nature was being represented without any reference to religious notions. The scientist was conforming with this tendency when, considering nature not only independently of God but even independently of man, he aimed at its 'objective' description or explanation. Yet it must be stressed that even for Newton the shell was important only because it stemmed from the great ocean of truth. Its observation was not yet a purpose in itself, but derived its significance from its connection with the ocean.

In the ensuing years the methods of Newtonian mechanics were applied successfully to ever greater realms of nature. Attempts were made to discover the details of natural processes by means of experiments, to observe them objectively and to understand the laws governing them. Scientists tried to formulate relationships mathematically, and to arrive at 'laws' which would hold without restriction in the entire cosmos. In this way they finally managed to harness the forces of nature to man's purposes. The magnificent development of mechanics in the eighteenth century, and of optics, thermodynamics and heat technology at the beginning of the nineteenth century are all evidence of the power of this approach.

Changes in the Meaning of the Word 'Nature'

Inasmuch as this kind of science was successful, it spread beyond the realm of daily experience into distant realms of nature, which could only be opened up properly by means of techniques which arose out of the development of science itself. Even in Newton's case, the decisive step had been his realization that the same laws of mechanics which governed the fall of a stone determined the motion of the moon about the earth; in other words, they could also be applied on a cosmic scale. In the period that followed, science began its victorious march on a broad front even into those distant realms of nature which could only be entered through technology, *i.e.*, by means of more or less complicated instruments. Astronomy, making use of ever better telescopes, conquered ever wider cosmic spaces. From the behaviour of matter during chemical changes, chemistry tried to fathom processes on the atomic scale. Experiments with the induction machine and the Voltaic cell provided the first common knowledge of electrical phenomena not yet understood. Thus, there took place a slow change in the significance of 'nature' as a subject for investigation by science. It became a collective concept for all those realms of experience into which man could penetrate by means of science and technology, regardless of whether or not they appeared as 'nature' to his immediate perception. Even the phrase 'description of nature' lost more and more of its original significance of a living and meaningful account of nature. Increasingly it became to mean the mathematical description of nature, *i.e.*, an accurate and concise yet comprehensive collection of data about relations that hold in nature.

This semi-conscious extension of the concept of nature must not yet be considered a basic departure from the original aims of science, since, even in this wider field, the crucial concepts were still the same as those of ordinary experience. In the nineteenth century nature still appeared as a set of laws in space and time in which man and man's intervention in nature could be ignored in principle, if not in practice.

Matter was thought of in terms of its mass, which remained constant through all changes, and which required forces to move it. Because, from the eighteenth century onwards, chemical experiments could be classified and explained by the atomic hypothesis of ancient times, it appeared reasonable to take over the view of ancient philosophy that atoms were the real substance, the immutable building-stones of matter. Just as in the philosophy of Democritus, the differences in material qualities were considered to be merely apparent; smell or colour, temperature or viscosity, were not actual qualities of matter but resulted from the interaction of matter and our senses, and had to be explained by the arrangements and movements of atoms, and by the effect of these arrangements on our minds. It is thus that there arose the over-simplified world-view of nineteenth-century materialism: atoms move in space and time as the real and immutable substances, and it is their arrangement and motion that create the colourful phenomena of the world of our senses.

The Crisis of the Materialist Conception

The first, but not yet very dangerous, incursion into this world-view took place in the second half of the last century with the development of the theory of electricity, in which not matter but fields of force were considered to be the real explanation. Interactions between fields of force without any matter to propagate the force were very much more difficult to understand than the materialist picture of atomic physics, and introduced an element of abstraction and a lack of clarity into what appeared otherwise to be so reasonable a world-view. Attempts were not lacking to return once more to the simpler concepts of materialist philosophy by way of the ether, which was supposed to be an elastic medium transmitting these fields of force; yet no such attempt had any real success. Even so, one could take comfort from the fact that changes in the fields of force could still be considered as processes in space and time, and that they could be described objectively, *i.e.*, without any reference to the manner in which they were observed, and thus in accordance with the generally held idealized view of laws of space and time. Furthermore, fields of force, *i.e.*, forces which can only be observed by their effect on atoms, could be considered as produced by atoms, and so as explaining atomic movements in some way. Thus atoms still remained as the actual essence, and between them there was empty space, real only inasmuch as it was a transmitter of fields of force.

In this world-view it did not matter overmuch that after the discovery of radio-activity at the end of the last century, the atoms of chemistry could no longer be considered as the ultimate indivisible building-stones of matter. These were now thought to consist of three kinds of basic units—the protons, neutrons and electrons of today. The practical consequences of this new knowledge have been the transmutation of elements and the rise of atomic physics, and they have thus become extremely important. Basically, however, nothing has been changed in principle by our acceptance of protons, neutrons and electrons as the smallest building-stones of matter, if we interpret these as the real essence. What is important for the materialistic world-view is simply the possibility that such small building-stones of elementary particles exist and that they may be considered the ultimate objective reality. Thus, the well-constructed world-view of the nineteenth and early twentieth centuries was preserved, and thanks to its simplicity it managed to retain its full power of conviction for a number of decades.

But in our century it is just in this sphere that fundamental changes have taken place in the basis of atomic physics which have made us abandon the world-view of ancient atomic philosophy. It has become clear that the desired objective reality of the elementary particles is too crude an oversimplification of what really happens, and that it must give way to very much more abstract conceptions. For if we wish to form a picture of the nature of these elementary particles, we can no longer ignore the physical processes through which we obtain our knowledge of them. While, in observing everyday objects, the physical process involved in making the observation plays a subsidiary role only, in the case of the smallest building particles of matter, every process of observation produces a large disturbance. We can no longer speak of the behaviour of the particle independently of the process of observation. As a final consequence, the natural laws formulated mathematically in quantum theory no longer deal with the elementary particles themselves but with our knowledge of them. Nor is it any longer possible to ask whether or not these particles exist in space and time objectively, since the only processes we can refer to as taking place are those which represent the interplay of particles with some other physical system, *e.g.*, a measuring instrument.

Thus, the objective reality of the elementary particles has been strangely dispersed, not into the fog of some new ill-defined or still unexplained conception of reality, but into the transparent clarity of a mathematics that no longer describes the behaviour of the elementary particles but only our knowledge of this behaviour. The atomic physicist has had to resign himself to the fact that his science is but a link in the infinite chain of man's argument with nature, *and that it cannot simply speak of nature 'in itself'*. Science always presupposes the existence of man and, as

Bohr has said, we must become conscious of the fact that we are not merely observers but also actors on the stage of life.

Technology

Intereffect of Technology and Science

Before we can speak of the general consequences of this new situation in modern physics, we must discuss the development of technology, which goes hand in hand with science and which is much more important for our practical life on earth. It is this very technology which has extended Western science over the whole world and has led to its occupying the central place in the thought of our time. In the developments of the last two hundred years, technology has always been both the starting point and consequence of natural science. It is the starting point since developments and clarifications of science often arise because of refinements in the means of observation alone. (We may remind the reader of the invention of the telescope, the microscope, or even the development of X-rays.) Technology is a consequence of science in that a technical exploitation of the forces of nature is generally possible only on the basis of a close understanding of a particular field.

Thus, starting in the eighteenth and the beginning of the nineteenth centuries, there was developed a technology which rested on the exploitation of mechanical processes. Often machines did nothing but imitate the action of our hands in spinning, weaving, lifting loads or forging large pieces of iron. Thus, this form of technology was at first merely the development and extension of old handicrafts, and outside-observers could understand it just as they had understood the old handicrafts themselves; everybody could grasp the underlying principles even though he might not be able to repeat the detailed manual skills. This characteristic of technology was not changed basically even by the introduction of the steam engine, although at this stage the extension of technology took place on an unprecedented scale, for the natural forces stored up in coal could now be placed at the service of man to take the place of manual labour.

However, a decisive change in the nature of technology did come about with the development of electro-technics in the second half of the last century. There was no longer a direct connection with the old handicrafts, since natural forces hardly known to man from his immediate experience of nature were being exploited. Even today, many people find something uncanny in electro-technics; it is felt to be incomprehensible even though it surrounds us everywhere. True, high-tension wires from which we are warned to keep away give us some intuitive ideas of the concept of fields of force that science has employed here, but at heart we feel that this is for us a closed chapter in the book of nature. Glimpses of the internal workings of a complicated electrical instrument are often felt to be as forbidding as observations of surgical operations.

Chemical technology might still have been considered as the continuation of old branches of handicrafts; we have but to think of dyeing, tanning and pharmacology. But the scope of chemical technology, as it has evolved since the turn of the century, no longer permits any comparisons with earlier conditions. Finally, atomic technology is exclusively concerned with the exploitation of natural forces to which there is no entry at all from the world of natural experience. True, this subject may one day become as commonplace as electro-technics is today, and even an integral part of everyday life. Yet the things by which we are daily surrounded do not thereby *belong to nature* in the original sense of the word. Perhaps the day will come when the many technical instruments will become as inescapable a part of ourselves as the snail's shell is to its occupant or as the web is to the spider. *But even these instruments would be a part of our own organism* rather than parts of external nature.

The Role of Technology in Man's Relationship with Nature

In all this, technology intervenes radically in the relationship of nature to man, radically changing his environment and thus bringing him face to face with the scientific aspect of the world. The claim of science, that it can reach into the whole universe by means of a method which, at a chosen moment, will isolate and illuminate details and thus advance from one relation to the next, is mirrored in technology, which progresses step by step to ever-new realms, changes our surroundings before our very eyes and thus stamps them with our image. Just as science subordinates every detailed question to the great task of understanding nature as a whole, so even the smallest technical advance serves the general aim of extending man's material powers. The value of this aim is questioned just as little as scientists question the value of an understanding of nature. Both aims become fused into the common-place slogan, 'knowledge is power'.

While this subordination to a single purpose can probably be proved to exist in every single technical process, the connection is often so indirect that it can hardly be considered a part of a conscious plan to reach an aim. Here technology no longer appears as the result of a conscious human effort to extend man's material powers, but rather as a large-scale biological process in which man's organic functions are increasingly transferred to his environment. In other words, we have here a biological process which, as such, is removed from man's control; for *while man can do what he wishes, he cannot will what he wishes.*

Science as a Part of the Interplay Between Man and Nature

Technology and Changes in our Way of Life

In this connection it has often been said that the far-reaching changes in our environment and in our way of life wrought by this technical age have also changed dangerously our ways of thinking, and that here lie the roots of the crises which have shaken our times and which, for instance, are also expressed in modern art. True, this objection is much older than modern technology and science, the use of implements going back to man's earliest beginnings. Thus, two and a half thousand years ago, the Chinese sage Chuang-Tzu spoke of the danger of the machine when he said:

'As Tzu-Gung was travelling through the regions north of the river Han, he saw an old man working in his vegetable garden. He had dug an irrigation ditch. The man would descend into the well, fetch up a vessel of water in his arms and pour it out into the ditch. While his efforts were tremendous the results appeared to be very meagre.

'Tzu-Gung said, "There is a way whereby you can irrigate a hundred ditches in one day, and whereby you can do much with little effort. Would you not like to hear of it?" Then the gardener stood up, looked at him and said, "And what would that be?"

'Tzu-Gung replied, "You take a wooden lever, weighted at the back and light in front. In this way you can bring up water so quickly that it just gushes out. This is called a draw-well."

'Then anger rose up in the old man's face, and he said, "I have heard my teacher say that whoever uses machines does all his work like a machine. He who does his work like a machine grows a heart like a machine, and he who carries the heart of a machine in his breast loses his simplicity. He who has lost his simplicity becomes unsure in the strivings of his soul. Uncertainty in the strivings of the soul is something which does not agree with honest sense. It is not that I do not know of such things; I am ashamed to use them." '

Clearly this ancient tale contains a great deal of wisdom, for 'uncertainty' in the 'strivings of the soul' is perhaps one of the aptest descriptions of man's condition in our modern crisis; technology, the machine, has spread through the world to a degree that our Chinese sage could not even have suspected. But, two thousand years have gone by and still man is creating the most beautiful works of art in the world, and that simplicity of heart of which the sage spoke has never been lost entirely. In the course of centuries, it may have been dimmed at times and have grown stronger at others, but always it has re-emerged in all its fruitfulness. After all, the rise of the human race is the result of the development of tools. Thus it cannot be technology itself that is the reason why our age has lost consciousness of so many values. We shall probably come closer to the truth if we blame the suddenness, and—compared with previous changes—the uncommonly fast development of technology in the last fifty years, for many of the difficulties. In contra-distinction to previous centuries this rapid change simply did not leave humanity time to get used to new conditions of life. This, however, fails to explain properly the entirely unprecedented nature of man's predicament.

Modern Man Confronts Only Himself

From the very start we have asked whether the changes in the foundations of modern science might not perhaps be considered as symptoms of shifts in the very basis of our existence, expressing themselves in various places simultaneously, be it in changes of our way of life, in our habits of thought, in external catastrophes, in wars or in revolutions. If, starting from the condition of modern science, we try to find out where the bases have started to shift, we get the impression that it would not be too crude an over-

NATURE IN CONTEMPORARY PHYSICS

simplification to say that *for the first time in the course of history modern man on this earth now confronts himself alone*, and that he no longer has partners or opponents.

In the first instance this applies very obviously to man's struggle with external dangers. Previously, man was threatened by wild animals, diseases, hunger, cold and other natural forces, and in this struggle every extension of technology meant a strengthening of man's position, *i.e.*, an advance. In our age, when the earth is becoming ever more densely populated, limitations of living possibilities, and thus the threat, are primarily due to other men who also are claiming their rights to the goods of this world. Here, extensions of technology need no longer lead to progress.

The phrase: 'Modern man confronts himself alone' is assuming an ever greater validity in this age of technology. In previous times man felt that he confronted nature alone. Nature populated by creatures of all kinds was a domain existing according to its own laws, to which he had somehow to adapt himself. In our age, however, we live in a world which man has changed so completely that in every sphere —whether we deal with the tools of daily life, whether we eat food which has been prepared by machines, or whether we travel in a countryside radically changed by man—we are always meeting man-made creations, so that in a sense we meet only ourselves. True, there are still parts of the earth where this process has by no means been concluded, but here also man's supremacy is bound to supervene sooner or later.

However, this new situation emerges most clearly in modern science itself where, as I have said previously, we can no longer consider 'in themselves' those building-stones of matter which we originally held to be the last objective reality. This is so because they defy all forms of objective location in space and time, and since basically it is always our knowledge of these particles alone which we can make the object of science. Thus the aim of research is no longer an understanding of atoms and their movements 'in themselves', *i.e.*, independently of the formulation of experimental problems. From the very start we are involved in the argument between nature and man in which science plays only a part, so that the common division of the world into subject and object, inner world and outer world, body and soul, is no longer adequate and leads us into difficulties. Thus even in science *the object of research is no longer nature itself, but man's investigation of nature*. Here, again, man confronts himself alone.

It is obviously the task of our age to come to terms with this new situation in every sphere of life, for only when we have been able to do so will we recover that 'certainty in the strivings of the soul' of which the Chinese sage has spoken. The road to this goal will be long and painful, and we do not know what Stations of the Cross we have yet to encounter on it, but if we are looking for some hints about the possible appearance of this road, it may be relevant to recall once more the example of the exact sciences.

The New Conception of Scientific Truth

In quantum theory, we accepted the situation described above once we could describe it mathematically and state clearly and categorically, and without any danger of logical contradictions, what would be the result of an experiment. Thus we accepted the new situation the moment we had removed all lack of clarity. True, the new mathematical formulae no longer describe nature itself but *our knowledge* of nature. We have had to forego the description of nature which for centuries was considered the obvious aim of all exact sciences. All we can say at present is that in the realm of modern atomic physics we have accepted this state of affairs because it describes our experience adequately. On the question of the philosophical interpretation of quantum theory opinions still differ, and occasionally we may hear the view that this new form of natural description is still unsatisfactory, since it fails to satisfy earlier ideals of what scientific truth ought to be and must thus be considered itself a symptom of the crisis of our times, and as by no means final.

In this connection it may be relevant to discuss the *concept of scientific truth* more generally, and to enquire what are the criteria which allow us to call scientific knowledge consistent and final. Let us begin with a purely external criterion. As long as any sphere of mental life advances continuously and without any inner break, those who work in this sphere will always pose detailed questions on what we may call problems of technique, whose solution is not a purpose in itself but whose value stems from the part they play in the larger framework which alone is important. Perhaps this is the reason why sculptors in the Middle Ages tried to give the best possible descriptions of folds in dresses, the solution of this particular problem being important since even the folds in the cloaks of the saints were a part of the great religious framework which was all the artist was really concerned about. Similarly, we find that modern science continues to pose specific problems and that work on them is the condition for an understanding of the larger framework. Even in the developments of the last fifty years particular problems constantly arose by themselves. They did not have to be looked for, and the aim was always that same great framework of natural laws. In this respect, and speaking purely from an external point of view, we can see no reason for any break in the continuity of the exact sciences.

With respect to the finality of the results, however, we must remind the reader that in the realm of the exact sciences there have always been final solutions for certain limited domains of experience. Thus, for instance, the questions posed by Newton's concept of mechanics found an answer valid for all time in Newton's law and in its mathematical consequences. True, these solutions are no more far-reaching than is the scope of Newton's mechanics and his particular way of posing problems. Thus, for instance, it was found that already in the theory of electricity an analysis using these concepts was no longer possible, and therefore in the investigation of this new domain of experience there emerged new systems of concepts leading to a final mathematical formulation of the laws of electricity.

Accordingly, in the exact sciences the word 'final' obviously means that there are always self-contained, mathematically representable, systems of concepts and laws applicable to certain realms of experience, in which realms they are always valid for the entire cosmos and cannot be changed or improved. Obviously, however, we cannot expect these concepts and laws to be suitable for the subsequent description of new realms of experience. It is only in this limited sense that quantum-theoretical concepts and laws can be considered as final, and only in this limited sense can it ever happen that scientific knowledge is finally formulated in mathematical or, for that matter, in any other language.

Similarly, many legal philosophies assume that while Law always exists, each new case generally involves a new discovery of law, and that the written law can be relevant only to limited realms of life and that it cannot be binding for ever. The exact sciences also start from the assumption that in the end it will always be possible to understand nature, even in every new field of experience, but that we may make no *a priori* assumptions about the meaning of the word 'understand'. In the sciences we find that the mathematical formulations of previous epochs are 'final' but by no means universal. It is because of this that it is impossible to base acts of faith, supposed to be binding for our behaviour in life, on our scientific understanding alone, since formulations of scientific knowledge apply only to a limited range of experience. Many modern creeds which claim that they are in fact not dealing with questions of faith, but are based on scientific knowledge, contain inner contradictions and rest on self-deception.

Nevertheless, we must not be misled into underestimating the firmness of the foundations of exact science. The concept of scientific truth, on which science is based, can apply to many different forms of knowledge. Thus, on it are based not only the sciences of the past centuries but modern atomic physics also, and this will make it clear why we can accept the fact that there are situations which no longer permit an objective understanding of natural processes, and yet use this realization to order our relationships with nature. When we speak of the picture of nature in the exact science of our age, we do not mean a picture of nature so much as a *picture of our relationships with nature*. The old division of the world into objective processes in space and time and the mind in which these processes are mirrored—in other words, the Cartesian difference between *res cogitans* and *res extensa*—is no longer a suitable starting point for our understanding of modern science. Science, we find, is now focused on the network of relationships between man and nature, on the framework which makes us as living beings dependent parts of nature, and which we as human beings have simultaneously made the object of our thoughts and actions. Science no longer confronts nature as an objective observer, but sees itself as an actor in this interplay between man and nature. The scientific method of analysing, explaining and classifying has become conscious of its limitations, which arise out of the fact that by its intervention science alters and refashions the object of investigation. In other words, method and object can no longer be separated. *The scientific world-view has ceased to be a scientific view in the true sense of the word.*

Consciousness of the Danger of Our Situation

However, by solving these paradoxes in a narrow scientific sphere, very little has been gained for the general situation of our age, in which, to repeat the above simplification, we suddenly and primarily confront ourselves alone. Hopes that the extension of man's material and spiritual powers would always spell progress are limited by this situation, if at first somewhat vaguely, and the dangers increase as the optimistic wave of faith in progress dashes against this limitation. Perhaps we might illustrate this kind of danger by means of an analogy.

In what appears to be its unlimited development of material powers, humanity finds itself in the position of a captain whose ship has been built so strongly of steel and iron that the magnetic needle of its compass no longer responds to anything but the iron structures of the ship; it no longer points north. The ship can no longer be steered to reach any goal, but will go round in circles, a victim of wind and currents. However, the danger persists only so long as the captain has not grasped that the compass is not responding to the magnetic forces of the earth. The moment he realizes that the danger is as good as half-removed; the captain who does not wish to sail in circles but wishes to reach a known or even unknown goal will find ways and means of determining the direction of his ship. He may use a modern compass which is not affected by the iron of the ship, or, as in the olden times, he may use the stars as his guides. Of course, he cannot order the stars to be visible at all times, and perhaps it is true that in our age only a few of them seem to be shining at all, but this one thing is clear: the very realization that faith in progress must have a limitation involves the wish to cease going in circles and to reach a goal instead.

As we become clearer about this limitation, the limitation itself may be considered to be the first foothold from which we may re-orientate ourselves.

Perhaps this analogy will help us in gaining a new hope that although these limitations affect us in some ways, they do not limit life itself. The space in which man develops as a spiritual being has more dimensions than the single one which it has occupied during the last centuries. This would imply that over longer periods of time, a conscious acceptance of this limitation might well lead to some equilibrium, where man's knowledge and creative forces will once again find themselves ranged spontaneously about their common centre.

2

ATOMIC PHYSICS AND CAUSAL LAW

Some of the most interesting general effects of modern atomic physics are the changes which it has brought about in the concept of natural laws. In the last few years many people have stated that modern atomic physics has abolished the law of cause and effect, or at least that it has shown the latter to be partially inoperative, and that we can no longer properly speak of processes determined by natural laws. Occasionally these statements simply assert that the principle of causality is no longer compatible with modern atomic theory. Such assertions are always vague if the concept of causality or natural law is not adequately defined. I should therefore like to say just a few words about the historical development of these concepts, before dealing with the relationship that prevailed between atomic physics and the principle of causality even long before the introduction of quantum theory. I shall then go on to discuss the consequences of quantum theory and the development of atomic physics in the last few years. Very little of this development has reached the public so far, but it looks very much as if this development also will have repercussions in the sphere of philosophy.

The Concept of Causality

The use of the concept of causality for describing the law of cause and effect is of relatively recent origin. In previous philosophies the word *causa* had a very much more general significance than it has today. Thus scholasticism, following Aristotle, spoke of four kinds of 'causes': the *causa formalis* which might be considered as the form or the spiritual essence of a thing, the *causa materialis* which referred to the matter of which the thing consisted, the *causa finalis* or the purpose for which the thing was created, and finally, the *causa efficiens*. Only the *causa efficiens* corresponds to what is meant by the word 'cause' today.

The transformations of *causa* into the modern concept of cause have taken place in the course of centuries, in close connection with the changes in man's conception of reality and with the creation of science at the beginning of the modern age. As material processes became more prominent in man's conception of reality, the word *causa* was used increasingly to refer to the particular material event which preceded, and had in some way caused, the event to be explained. Thus even Kant, who frequently did what at root amounted to drawing philosophic consequences from the developments in science since Newton's time, already used the word 'causality' in a nineteenth-century sense. When we experience an event we always assume that there was another event preceding it from which the second has followed according to some law. Thus the concept of causality became narrowed down, finally, to refer to our belief that events in nature are uniquely determined, or, in other words, that an exact knowledge of nature or some part of it would suffice, at least in principle, to determine the future. Newton's physics was so constructed that the future motion of a system could be calculated from its particular state at a given time. The idea that nature really was like this was perhaps enunciated most generally and most lucidly by Laplace when he spoke of a demon, who at a given time, by knowing the position and motion of every atom, would be capable of predicting the entire future of the world. When the word 'causality' is interpreted in this very narrow sense, we speak of 'determinism', by which we mean that there are immutable natural laws that uniquely determine the future state of any system from its present state.

Statistical Laws

From its very beginnings atomic physics evolved concepts which do not really fit this picture. True, they do not contradict it basically, but the approach of atomic physics was by its very character different from that of determinism. Even in the ancient atomic theory of Democritus and Leucippus it was assumed that large-scale processes were the results of many irregular processes on a small scale. That this is basically the case is illustrated by innumerable examples in everyday life. Thus, a farmer need only know that a cloud has condensed and watered his fields. He does not bother about the path of each individual drop of rain. To give another example, we know precisely what is meant by the word 'granite', even when we are ignorant of the form, colour, and chemical composition of each small constituent crystal. Thus we always use concepts which describe behaviour on the large scale without in the least bothering about the individual processes that take place on the small scale.

This notion of the statistical combination of many small individual events was already used in ancient atomic theory as the basis for an explanation of the world, and was generalized in the concept that all the sensory qualities of matter were indirectly caused by the position and movements of the atoms. Thus Democritus wrote that things only *appeared* to be sweet or bitter, and only appeared to have colour, for in reality there existed only atoms and empty space. Now, if the processes which we can observe with our senses are thought to arise out of the interactions of many small individual processes, we must needs conclude that all natural laws may be considered to be only statistical laws. True, even statistical law can lead to statements with so high a degree of probability that they are almost certain, but there can always be exceptions in principle. The concept of statistical law is frequently thought to be contradictory. Thus it is contended that while it is possible to look upon natural processes either as determined by laws, or else as running their course without any order whatever, we cannot form any picture of processes obeying statistical laws.

Yet we must remind the reader that in everyday life all of us encounter statistical laws with every step we take, and make these laws the basis of our practical actions. Thus, when an engineer is constructing a dam he always bases his calculations on the average yearly rainfall, although he cannot have the faintest idea when it will rain and how much of it at a time. When speaking of statistical laws we generally mean that a particular physical system is known incompletely. The most common example is the throw of dice. Since no one side of a die is heavier than any other, and since it is thus impossible to predict which side will turn up, we can assume that in a large number of throws precisely one in six will turn up with five dots.

From the very beginning of modern times attempts have been made to explain both qualitatively and also quantitatively the behaviour of matter

ATOMIC PHYSICS AND CAUSAL LAW

through the statistical behaviour of atoms. Robert Boyle demonstrated that we could understand the relations between the pressure and the volume of a gas if we looked upon pressure as the many thrusts of the individual atoms on the walls of the vessel. Similarly thermodynamical phenomena have been explained by the assumption that atoms move more violently in a hot body than in a cold one. This statement could be given a quantitative mathematical formulation and in this way the laws of heat could be understood.

This application of the concept of statistical laws was finally formulated in the second half of the last century as the so-called *statistical mechanics*. In this theory, which is based on Newton's mechanics, the consequences that spring from an incomplete knowledge of a complicated mechanical system are investigated. Thus in principle it is not a renunciation of determinism. While it is held that the details of events are fully determined according to the laws of Newton's mechanics, the condition is added that the mechanical properties of the *system* are not fully known.

Gibbs and Boltzmann managed to formulate this kind of incomplete knowledge mathematically, and Gibbs was able to demonstrate that, in particular, our conception of temperature is closely related to the incompleteness of our knowledge. When we know the temperature of a particular system, it means that the system must be considered to be only one out of a whole set of systems. This set of systems can be described accurately by mathematics, but not the particular system with which we are concerned. With this Gibbs had half-unconsciously taken a step which later on was to have the most important consequences. Gibbs was the first to introduce a physical concept which can only be applied to an object when our knowledge of the object is incomplete. If for instance the motion and position of each molecule in a gas were known, then it would be pointless to continue speaking of the temperature of the gas. The concept of temperature can only be used meaningfully when the system is not fully known and we wish to derive statistical conclusions from our incomplete knowledge.

The Statistical Character of Quantum Theory

Although the discoveries of Gibbs and Boltzmann made an incomplete knowledge of a system part of the formulation of physical laws, nevertheless determinism was still present in principle until Max Planck's famous discovery ushered in quantum theory. Planck, in his work on the theory of radiation, had originally encountered an element of uncertainty in radiation phenomena. He had shown that a radiating atom does not deliver up its energy continuously, but discreetly in bundles. This assumption of a discontinuous and pulse-like transfer of energy, like every other notion of atomic theory, leads us once more to the idea that the emission of radiation is a statistical phenomenon. However, it took two and a half decades before it became clear *that quantum theory actually forces us to formulate these laws precisely as statistical laws* and to depart radically from determinism. Since the work of Einstein, Bohr and Sommerfeld, Planck's theory has proved to be the key with which the door to the entire sphere of atomic physics could be opened. Chemical processes could be explained by means of the Rutherford-Bohr atomic model, and since then, chemistry, physics and astrophysics have been fused into unity. With the mathematical formulation of quantum-theoretical laws pure determinism had to be abandoned.

Since I cannot speak of the mathematical methods here, I should merely like to mention some aspects of the strange situation confronting the physicist in atomic physics.

We can express the departure from previous forms of physics by means of the so-called uncertainty relations. It was discovered that it was impossible to describe simultaneously both the position and the velocity of an atomic particle with any prescribed degree of accuracy. We can either measure the position very accurately—when the action of the instrument used for the observation obscures our knowledge of the velocity, or we can make accurate measurements of the velocity and forego knowledge of the position. The product of the two uncertainties can never be less than Planck's constant. This formulation makes it quite clear that we cannot make much headway with the concepts of Newtonian mechanics, since in the calculation of a mechanical process it is essential to know simultaneously the position and velocity at a particular moment, and this is precisely what quantum theory considers to be impossible.

Another formulation is that of Niels Bohr, who introduced the *concept of complementarity*. By this he means that the different intuitive pictures which we use to describe atomic systems, although fully adequate for given experiments, are nevertheless mutually exclusive. Thus, for instance, the Bohr atom can be described as a small-scale planetary system, having a central atomic nucleus about which the external electrons revolve. For other experiments, however, it might be more convenient to imagine that the atomic nucleus is surrounded by a system of stationary waves whose frequency is characteristic of the radiation emanating from the atom. Finally, we can consider the atom chemically. We can calculate its heat of reaction when it becomes fused with other atoms, but in that case we cannot simultaneously describe the motion of the electrons. Each picture is legitimate when used in the right place, but the different pictures are contradictory and therefore we call them mutually complementary. The uncertainty that is attached to each of them is expressed by the uncertainty relation, which is sufficient for avoiding logical contradiction between the different pictures.

Even without entering into the mathematics of quantum theory these brief comments might have helped us to realize *that the incomplete knowledge of a system must be an essential part of every formulation in quantum theory*. Quantum theoretical laws must be of a statistical kind. To give an example: we know that the radium atom emits alpha-radiation. Quantum theory can give us an indication of the probability that the alpha-particle will leave the nucleus in unit time, but it cannot predict at what precise point in time the emission will occur, for this is uncertain in principle. We cannot even assume that new laws still to be discovered will allow us to determine this precise point in time; were this possible the alpha-particle could not also be considered to behave as a wave leaving the atomic nucleus, a fact which we can prove experimentally. The various experiments proving both the wave nature and also the particle nature of atomic matter create a paradox which forces us to devise a formulation of statistical laws.

In large-scale processes this statistical aspect of atomic physics does not arise, generally because statistical laws for large-scale processes lead to such high probabilities that to all intents and purposes we can speak of the processes as determined. Frequently, however, there arise cases in which a large-scale process depends on the behaviour of one or of a few atoms alone. In that case, the large-scale process also can only be predicted statistically. I should like to illustrate this by means of a well-known but unhappy example, that of the atom bomb. In an ordinary bomb the strength of the explosion can be predicted from the mass of the explosive material and its chemical composition. In the atom bomb we can still indicate an upper and a lower limit of the strength of the explosion but we cannot make exact calculations of this strength in advance. This is impossible in principle since it depends on the behaviour of only a few atoms at the moment of firing.

Similarly, there may be biological processes—and Jordan, especially, has drawn our attention to this—in which large-scale events are set off by processes in individual atoms; this would appear to be the case particularly in the mutation of genes during hereditary processes. These two examples were meant to illustrate the practical consequences of the statistical character of quantum theory. This development, too, was concluded over two decades ago, and we cannot possibly assume that the future will see any basic changes in this field.

The History of More Recent Atomic Physics

However, very recently, a new point of view has been added to the problem of causality, which, as I have said at the beginning, stems from the latest developments in atomic physics. The questions of paramount interest today arose logically out of the progress of atomic physics during the last two hundred years, and therefore I must once more deal with the history of fairly recent atomic physics. At the beginning of the modern age the atomic concept was closely linked with that of chemical elements. An element was characterized by the fact that it could not be decomposed further by chemical means, and by the fact that a particular kind of atom belonged to a particular element. Thus, pure carbon consists of carbon atoms only, and pure iron of iron atoms only, and one was forced to assume that there were just as many sorts of atoms as there were chemical elements. Since ninety-two different kinds of chemical elements were eventually known, one assumed the existence of ninety-two kinds of atoms.

However, this conception is not satisfying if we approach the problem from the basic premises of atomic theory. Originally atoms were introduced to explain matter qualitatively through their movements and structure. This conception can have a true explanatory value only if all the atoms are equal, or if there are but a few kinds of atoms—in other words, if the atoms themselves have no qualities at all. If, now, one is forced to assume the existence of ninety-two qualitatively different atoms, this is no great gain over the simple assertion that there are things which differ qualitatively. Thus the assumption of ninety-two different basic particles has long been felt to be unsatisfactory and attempts were made to reduce these ninety-two kinds of atoms to a smaller number of elementary particles. Quite early on it was thought that chemical atoms themselves might well be composed of a very small number of basic building-stones. After all, even the oldest attempts to change one chemical substance into another must have been based on the assumption that in the final analysis all matter was one.

The last fifty years have shown that all chemical atoms are composed of only three basic building-stones, which we call protons, neutrons and electrons. The atomic nucleus consists of protons and neutrons and is surrounded by a number of electrons. Thus, for instance, the nucleus of the carbon atom consists of six protons and six neutrons and is surrounded by six electrons at a relatively great distance. Thanks to the development of nuclear physics in the 1930's we now have these three different kinds of particles, instead of the ninety-two different kinds of atoms, and in this respect atomic theory has followed the very path which its basic assumptions had suggested. When it became clear that all the atoms were composed of only three kinds of basic building-stones, there arose the practical possibility of changing chemical elements into one another. We know that this physical possibility was soon to be followed by its technical realization. Since Otto Hahn's *discovery of the fission of uranium* in 1938, and following the technical developments which sprang from it, we can now transform elements even on a large scale.

However, during the last two decades, the picture has once more become a little confused. In addition to the three elementary particles already mentioned, the proton, neutron and electron, new elementary particles were discovered in the thirties and during the last few years their number has increased most disturbingly. In contrast to the three basic building-stones, these new particles are always unstable and have very short lifetimes. Of these so-called *mesons*, one type has a lifetime of about a millionth of a second, another lives for only one hundredth part of that time, and a third, which has no electrical charge, for only a hundred billionth of a second. However, apart from their instability, these three new elementary particles behave very similarly to the three stable building-stones of matter. At first glance it looks as if, once more, we were forced to assume the existence of a great number of qualitatively different particles, which would be most unsatisfactory in view of the basic assumptions of atomic physics. However, during experiments in the last few years, it has become clear that these elementary particles can change into one another during their collisions, with great changes of energy. When two elementary particles collide with a great energy of motion, new elementary particles are created and the original particles, together with their energy, are changed into new matter.

This state of affairs is best described by saying that all particles are basically nothing but different stationary states of one and the same stuff. Thus even the three basic building-stones have become reduced to a single one. *There is only one kind of matter but it can exist in different discrete stationary conditions.* Some of these conditions, *i.e.*, protons, neutrons and electrons, are stable while many others are unstable.

Relativity Theory and the Dissolution of Determinism

Although the experimental results of the last few years leave us hardly any doubt that atomic physics will develop in this particular direction, no one has yet found a mathematical formulation of the laws governing the formation of the elementary particles. This is the very problem on which atomic physicists are working at the moment, both experimentally by discovering new particles and investigating their properties, and theoretically through efforts to correlate the laws governing the properties of the elementary particles and to formulate these mathematically.

In their efforts physicists have met with difficulties in the concept of time. When dealing with collisions of high-energy elementary particles we must consider the space-time structure of special relativity theory. This space-time structure was not very important in the quantum theory of the atomic shell, since in it the electrons move relatively slowly. Now, however, we are dealing with elementary particles which move almost with the velocity of light, and whose behaviour can therefore only be described with the help of relativity theory. Fifty years ago Einstein discovered that the structure of space and time was not quite as simple as we imagine it to be in everyday life. If we describe all those events as past of which, at least in principle, we can obtain some knowledge, and as future all those events on which, at least in principle, we can still have some influence, then according to our naïve conception we believe that between these two types of events there is but one infinitely short moment which we call the present. This was just the conception on which Newton had based his mechanics. Since Einstein's discovery in 1905, we know that between what I have just called 'future' and 'past' there exists an interval whose extension in time depends on the distance in space between an event and its observer. Thus, the present is not limited to an infinitely short moment in time.

Relativity theory assumes that in principle no effect can be propagated faster than the velocity of light. Now this trend in relativity theory leads to difficulties in connection with the uncertainty relations of quantum theory. According to relativity theory the only effects possible are in that part of space-time limited by the so-called light-cone, *i.e.*, those points in space-time which can be reached by a lightwave emanating from the effective point. This region in space-time is thus—and this must be stressed—very strictly limited. On the other hand, we have found that in quantum theory a clear determination of position—in other words, a sharp delimitation of space—presupposes an infinite uncertainty of velocity and thus also of momentum and energy. This state of affairs has as its practical consequence the fact that in attempting to arrive at a mathematical formulation of the interactions of the elementary particles, we shall always encounter infinite values for energy and momentum, preventing a satisfactory mathematical statement.

In recent years many investigations have been concerned with these difficulties, but so far no one has been able to arrive at a satisfactory solution. The only consolation is the assumption that in very small regions of space-time of the order of magnitude of the elementary particles, the notions of space and time become unclear, *i.e.*, in very small intervals even the concepts 'earlier' and 'later' can no longer be properly defined. Of course nothing is altered in space-time on the large scale, but we must bear in mind the possibility that experiment may well prove that small-scale space-time processes may run in reverse to the causal sequence. It is here that the latest developments of atomic physics once again come up against the question of causal laws. It is too early to say whether this will lead to new paradoxes, or that new deviations from causal laws will appear. Perhaps our attempts to formulate the laws of elementary particles mathematically will lead to new possibilities enabling us to avoid these difficulties. However, even now it has become quite certain that the latest developments in atomic physics will once more have repercussions in the sphere of philosophy. The final answer to the questions we have just posed will only come with the mathematical formulation of the natural laws governing the behaviour of elementary particles; when, for instance, we shall know why it is that the proton happens to be precisely 1836 times as heavy as the electron.

All this will have made it clear that atomic physics has moved ever further away from the concept of determinism. At first, from the very beginnings of atomic theory, the laws governing large-scale processes were looked upon as statistical laws. Although determinism was in this way preserved in principle, in practice it meant that we took account of our incomplete knowledge of physical systems. Then, in the first half of our century, incomplete knowledge of atomic systems was recognized as being a part of theory in principle. Finally, in very recent years we have come to realize that on a small space-time scale the concept of a sequence in time has become problematical, and we cannot even tell how this riddle will ever be solved.

3

CLASSICAL EDUCATION, SCIENCE AND THE WEST

The Traditional Reasons for the Defence of a Classical Education

MANY people have asked whether a classical education is not too theoretical or unworldly, and whether in our age of technology and science a more practical education would not be much more suited to equip us for life. This bears directly on the frequently discussed question of the relationship between the humanities and contemporary science. I cannot deal with this question fundamentally, for I am not a pedagogue nor have I been overmuch concerned with educational problems. I can, however, try to recall my own experiences, since I myself had a classical education and later on devoted most of my work to science.

What are the arguments that defenders of the humanities have produced, time and again, in favour of concentrating on ancient languages and ancient history? In the first place, they rightly point to the fact that our whole cultural life, our actions, our thoughts and our feelings, are steeped in the spiritual roots of the West, *i.e.*, in that attitude of mind which in ancient times was initiated by Greek art, Greek poetry and Greek philosophy. With the rise of Christianity and the formation of the Church great changes took place, and finally, at the end of the Middle Ages, there occurred the tremendous fusion of Christian piety with the Greek spirit of enquiry, and the world, as God's world, was radically altered by voyages of discovery, by science and by technology. In every sphere of modern life examination of the root of things, whether methodological, historical, or philosophical, brings us up against the concepts of antiquity and Christianity. Thus we may say in favour of a classical education that it is always a good thing to know these roots, even if they may not always be of practical use.

Second, we must stress the fact that the whole strength of our Western culture is derived, and always has been derived, from the close relationship between the way in which we pose our questions and the way in which we act. In the sphere of practical action other people and other cultural groups were just as experienced as the Greeks, but what always distinguished Greek thought from that of all other peoples was its ability to change the questions it asked into questions of principle and thus to arrive at new points of view, bringing order into the colourful kaleidoscope of experience and making it accessible to human thought. It is this link between the posing of questions of principle and practical action which has distinguished Greek thought from all others, and which during the rise of the West at the time of the Renaissance, the turning-point in our history, was responsible for the rise of modern science and technology. Whoever delves into the philosophy of the Greeks will encounter at every step this ability to pose questions of principle, and thus by reading the Greeks he can become practised in the use of the strongest mental tool produced by Western thought. Hence, in this respect, we can fairly say that a classical education teaches us something very important.

Finally, it is justly said that a concern with antiquity gives us a judgement in which spiritual values are prized higher than material ones. It is precisely in Greek thought, and in all the traces of it that we have inherited, that the pre-eminence of the spirit clearly emerges. True, people of today might take exception to just this fact, for they might say that our age has demonstrated that only material power, raw materials and industry are important, and that physical power is stronger than spiritual might. It would follow that it is not in the spirit of the times to teach our children to attach greater importance to spiritual than to material values.

In this connection, I am reminded of a conversation which I had some thirty years ago in the forecourt of our University. At that time Munich was in the throes of a revolution. The inner town was still occupied by the Communists, and I, then seventeen years old, had been assigned with some school comrades as auxiliaries to a military unit which had its headquarters opposite the University, in the Theological Seminary. Why all this happened is no longer quite clear to me, but it is probable that we found these weeks of playing at soldiers to be a very pleasant interruption of our lessons at the Maximilian Gymnasium. In the Ludwig Strasse there was occasional, if not very heavy, shooting. At noon we fetched our meals from a field-kitchen in the University courtyard. On one such occasion we had a discussion with a theology student on the question whether these struggles in Munich had any meaning, and one of us younger ones said emphatically that questions of power could never be settled by spiritual means, by speeches or by writing; only force could lead to a real settlement of our conflicts with others.

The theology student replied that in the final analysis even the question of what was meant by 'we' and 'the others', and what distinguished the two, would obviously lead to a purely spiritual decision, and that in all probability we should have gained a great deal if we could settle this question more reasonably than by the usual method. We could hardly object to this. Once the arrow has left the bow, it flies on its path, and only a stronger force can divert it; but its original direction was determined by him who aimed, and without the presence of a spiritual being with an aim it would never have been able even to start on its flight. In this regard we could do far worse than teach our youth not to rate spiritual values too low.

The Mathematical Description of Nature

However, I have strayed too far from my proper theme, and I must revert to my first real encounter with science at the Maximilian Gymnasium in Munich, since, after all, I am speaking of the relation between science and a classical education. Most schoolboys are introduced to technology and science when they begin to play with apparatus. Emulating the example of a fellow pupil, or perhaps because of a present received at Christmas—or even through school lessons—they begin to have a desire to handle small engines, and perhaps even to build one. This is precisely what I did with great enthusiasm during the first five years of my life at high-school. This activity would probably have remained a mere game and would not have led me to real science, if another event had not occurred.

At the time, we were being taught the basic axioms of geometry. At first, I felt this to be very dry stuff; triangles and rectangles do not kindle one's imagination as much as do flowers and poems. But then our outstanding mathematics teacher, Wolff by name, introduced us to the idea that one could formulate generally valid propositions from these figures, and that some results, quite apart from their demonstrable geometric properties, could also be proved mathematically. The thought that mathematics somehow corresponded to the structures of our experience struck me as remarkably strange and exciting.

What had happened to me was what happens only too rarely with the intellectual gifts we are handed at school, for school lessons generally allow the different landscapes of the world of the mind to pass by our eyes, without quite letting us become at home in them. According to the teacher's abilities these landscapes are illuminated more or less brightly, and we remember the pictures for a shorter or a longer time. However, very occasionally, an object that has thus come into our field of view will suddenly begin to shine in its own light, first dimly and vaguely, then ever more brightly, until finally it will glow through our entire mind, spill over to other subjects and eventually become an important part of our own life. This happened in my case with the realization that mathematics fitted the things of our experience, a realization which, as I learnt at school, had already been gained by the Greeks, by Pythagoras and by Euclid.

At first, stimulated by Herr Wolff's lessons, I tried out this application of mathematics for myself, and

I found that this game between mathematics and immediate perception was at least as amusing as most other games. Later on, I discovered that geometry alone was no longer adequate for this mathematical game which had given me so much pleasure. From some books I gleaned that the behaviour of quite a few of my home-made instruments could also be described mathematically and I now began to read voraciously in somewhat primitive mathematical textbooks, in order to acquire the mathematics needed for the description of physical laws, *i.e.*, the differential and integral calculus. In all this I saw the achievements of modern times, of Newton and his successors, as the immediate consequence of the efforts of the Greek mathematicians or philosophers, and never once did it occur to me to consider the science and technology of our times as belonging to a world basically different from that of the philosophy of Pythagoras or Euclid.

Although, in my youthful ignorance, I was not fully aware of it, this enjoyment of the mathematical description of nature had introduced me to the basic trait of all Western thought, namely, to the interrelationship between the way in which we pose questions and the way in which we act. Mathematics is, so to speak, the language in which the questions are posed and answered, but the questions themselves are concerned with processes in the practical material world; thus, geometry, for instance, was designed for measuring agricultural land. Because of all this, I remained far more interested in mathematics than in science or apparatus during most of my life at school, and it was only in the two upper classes that I acquired a special liking for physics—oddly enough because of a fortuitous encounter with a fragment of the modern physics.

Atoms and Classical Education

At that time we were using a rather good textbook of physics in which, quite understandably, modern physics was treated in a somewhat off-hand manner. However, the last few pages of the book dealt briefly with atoms and I distinctly remember an illustration depicting a large number of them. The picture was obviously meant to represent the state of a gas on a large scale. Some of the atoms were clustered in groups and were connected by means of hooks and eyes supposedly representing their chemical bonds. On the other hand, the text itself stated that according to the concepts of the Greek philosophers atoms were the smallest indivisible building-stones of matter. I was greatly put off by this illustration, and I was enraged by the fact that such idiotic things should be presented in a textbook of physics, for I thought that if atoms were indeed such crude structures as this book made out, if their structure was complicated enough for them to have hooks and

CLASSICAL EDUCATION 59

eyes, then they could not possibly be the smallest indivisible building-stones of matter.

In my criticisms I was supported by a friend from my youth club with whom I had gone on many hiking expeditions, and who was much more interested in philosophy than I was. This friend, who had read some essays on atomic theory in ancient philosophy, had also unexpectedly come across a textbook of modern atomic physics (I believe it was Sommerfeld's *Atomic Structure and Spectral Lines*) where he had seen visual models of atoms. This had led him to the firm conviction that the whole of modern atomic physics was false, and he tried to convince me that he was right. At that time our judgements were obviously very much rasher and more dogmatic than they are today. I had to agree with him that these visual models of atoms were indeed false, but I reserved the right to look for the mistakes in the illustrations rather than in the theory.

In any case, I had gained the wish to become better acquainted with the case for atomic physics, and here another accident came to my aid. At the time we had just started reading one of Plato's *Dialogues*, but school lessons were irregular. I have already told how I, as a young boy, had been a member of a military unit during the Munich revolution and that we had been stationed in the Theological Seminary opposite the University. We had no rigid plan of work at the time—far from it. The danger of lounging about was very much greater than that of over-exertion. In addition, we had to be prepared to be called even at night, and thus we were without any control by parents or teachers.

It was then July, 1919 (a warm summer), and there were hardly any military duties, particularly in the early mornings. Thus it came about that frequently, shortly after sunrise, I would withdraw on to the roof of the Theological Seminary and lie down there to warm myself in the sun, any old book in my hand; or I would sit on the edge of the roof and watch the day beginning in the Ludwig Strasse.

On one such occasion, it occurred to me to take a volume of Plato on to the roof, for I wanted to read something different from the books we were supposed to study in school. With my somewhat modest Greek knowledge, I came upon the dialogue called *Timaeus*, where for the first time and from the original source I read something about Greek atomic philosophy. This lecture made the basic thoughts of atomic theory much clearer to me than they had been; or at least, I believed that now I had an inkling of the reasons that had in the first place caused Greek philosophy to conceive of these smallest indivisible building-stones of matter. True, I did not feel that Plato's thesis in *Timaeus*—i.e., that atoms are uniform bodies—was fully convincing, but at least I was happy to learn that they did not have hooks and eyes. In any case, at that time I was gaining the growing

CLASSICAL EDUCATION

conviction that one could hardly make progress in modern atomic physics without a knowledge of Greek natural philosophy, and I thought that our illustrator of the atomic model would have done well to make a careful study of Plato before producing his particular illustration.

Thus, without properly knowing how, I had become acquainted with that great thought of Greek natural philosophy which links antiquity with modern times and which only came to full fruition at the time of the Renaissance. This trend in Greek philosophy, typified by the atomic theory of Leucippus and Democritus, used to be described as 'materialism'. Historically this is a correct description, but today it is easily misunderstood, since the word 'materialism' was given a very one-sided bias in the nineteenth century by no means in accordance with its meaning in Greek natural philosophy. We can avoid this false interpretation of ancient atomic theory if we remember that the first modern investigator to return to the atomic theory in the seventeenth century was the theologian and philosopher Gassendi, who, we may be sure, did not use the theory in order to combat the dogma of the Christian religion; indeed, even for Democritus atoms were merely the letters with which we could record the events of the world, but not their content. In contradistinction, nineteenth-century materialism was developed from thoughts of quite a different kind, thoughts which are characteristic of the modern age and are rooted in the division of the world into separate material and spiritual realities, as proposed by Descartes.

Science and Classical Education

We have seen that the great stream of science and technology of modern times springs from two sources in the fields of ancient philosophy. Although many other tributaries have flowed into this stream, and have helped to swell its current, the origins have always continued to make themselves felt. Because of all this the sciences cannot but benefit from classical studies. People who are concerned with the more practical schooling of youth for their struggles in later life will continue to assert that the knowledge of this spiritual foundation has little relevance for practical activities, and that they should rather acquire the necessities of modern life: modern languages, technical methods, accounting and commercial practice. These (they say) will set the youngsters on their feet, but a classical education, being, so to speak, merely of decorative value, is a luxury which only those few can afford for whom fate has made the struggle for life less exacting. Perhaps this is true for the many people who will do nothing in their later lives but carry on a purely practical business, and who themselves will have no wish to influence the spiritual climate of their age. Those, however, who find this inadequate, and wish to get to the root of things in their chosen vocation, whether it be in technology or medicine, are bound sooner or later to encounter the sources of antiquity, and their own work can only benefit if they have learnt from the Greeks how to discipline their thoughts and how to pose questions of principle. I believe that in the work of Max Planck, for instance, we can clearly see that his thought was influenced and made fruitful by his classical schooling.

Perhaps I may here cite yet another personal experience which occurred three years after I had left school. While a student at Göttingen, I discussed with a fellow student the problem of the model of the atom that I had found so disturbing while still at school. This question was obviously the basis of the puzzling phenomena of spectroscopy which were still unsolved at that time. This friend defended perceptual models, and he believed that all that was needed was to enrol the help of modern technology in the construction of a microscope with a very great resolving power—for example, one employing gamma-rays instead of ordinary light. We should then be able to see the structure of the atom, and so my objections to perceptual models would finally be dispelled.

This argument disquieted me deeply. I was afraid that this imaginary microscope might well reveal the hooks and eyes of my physics textbook, and once again I had to resolve the apparent contradiction between this envisaged experiment and the basic conceptions of Greek philosophy. Here the education in disciplined thought that we had received at school was to help me a great deal, and make me wary of accepting unproved solutions. In this I was greatly helped also by what little acquaintance with Greek natural philosophy I had made at that time.

In contemporary discussions about the value of a classical education one can no longer maintain that the relationship between natural philosophy and modern atomic physics is a unique case. For even if we rarely meet such questions of principle in technology science or medicine, these disciplines are basically connected with atomic physics and thus, in the final analysis, lead to similar questions of principle. Chemical structure is explained on the basis of atomic physics. Modern astronomy is connected with it most closely, and can hardly make any progress without it. Even in biology, many bridges are being built towards atomic physics. The connections between the different branches of science have become much more obvious in the last decades than at any previous time. There are many signs of their common origin, which, in the final analysis, must be sought somewhere in the thought of antiquity.

Faith in Our Task

With this conclusion I have almost returned to the point from which I started. At the root of all Western culture there is this close connection between our way of posing questions of principle and our actions; this we owe to the Greeks. Even today the whole force of our culture rests on this connection. From it springs all our progress, and in this sense a declaration of faith in a classical education is an avowal for the West and for its culture.

However, do we still have a right to this faith when the West has lost so terribly in power and prestige in the last decades? Our answer is that all this does not involve questions of right, but questions of *will*. For the activity of the West does not stem from theoretical insights—our ancestors did not base their actions on theories—but from quite a different origin. What is, and always has been, our mainspring, is faith. By faith I do not mean only the Christian faith in a God-given and meaningful framework of the world, but simply faith in our task in this world. Here, faith obviously does not mean that we hold this or that to be true. If I have faith, it means that I have decided to do something and am willing to stake my life on it. When Columbus started on his first voyage into the West, he believed that the earth was round and small enough to be circumnavigated. He did not think that this was right in theory alone, but he staked his whole existence on it.

In a recent discussion of this aspect of European history, Freyer has rightly referred to the old saying: '*Credo ut intellegam*'—'I believe so that I may understand'. In extending the application of this idea to the voyages of discovery, Freyer introduced an intermediate term: '*Credo, ut agam; ago, ut intellegam*'—'I believe so that I may act; I act so that I may understand'. This saying is relevant not only to the first voyages round the world, it is relevant to the whole of Western science, and also to the whole mission of the West. It includes classical education as well as science. And there is no need to be over-modest: one half of the modern world—the West—has gained immeasurable power by applying the Western idea of controlling and exploiting natural resources by means of science. The other half of the world, the East, is held together by its faith in the scientific theories of a European philosopher and political economist. Nobody knows what the future will hold and what spiritual forces will govern the world, but our first step is always an act of faith in something and a wish for something.

We wish that spiritual life may once again blossom here, that here in Europe thoughts may continue to grow and shape the face of the world. We stake our existence on this, and in so far as we remember our origins, and recover the harmonious interplay of Western influences, we shall make the external conditions of life in the West happier than they have been for fifty years. We wish that, despite all outer confusion, our youth will grow up in the spiritual climate of the West, and so draw on those sources of vitality which have sustained our continent for more than two thousand years. Let us not worry about the detailed ways in which this might be brought about. It does not matter whether we prefer a classical or a scientific education. What alone matters is our unshakable faith in the West.

Historical Sources

4
THE BEGINNINGS OF MODERN SCIENCE

In this essay I have tried to give a brief sketch of the philosophical problems that have arisen out of changes in the world-view of physics and science. These problems endow historical relations with a quite special significance; and the reader is now to be introduced to some original sources that he may himself follow these changes in the attitude of science.

In so slender a volume it is obviously impossible to quote from even a reasonably adequate list of sources, and I have merely endeavoured to illustrate some of the essential turning points, thus hoping that my preceding arguments may be more fully appreciated.

Johannes Kepler
(1571–1630)

Science at the end of the sixteenth and at the beginning of the seventeenth centuries was still largely influenced by the world-view of the Middle Ages, which saw nature primarily as created by God.

'There are three things, above all, whose causes,

that is why they are such and no other, I have investigated unceasingly, namely the number, magnitude and motion of the orbits. I was encouraged in my daring attempts by that beautiful harmony of the bodies at rest, namely the Sun, the fixed stars and the intervening space containing God the Father, God the Son and God the Holy Ghost.' Thus reads the 'Preface to the Reader' in the *Mysterium Cosmographicum* by Johannes Kepler. To give praise to God, we must read in the book of nature. In creating the world He used order and law, and endowed man not only with the senses but also with a mind, so that from the existence of the things that he can observe with his own eyes, he might conclude as to the causes of their being and becoming. There exists a full correspondence between man's abilities and the reality of creation, mirroring the all-embracing harmony. 'I believe that the causes of most things in the world are to be found in God's love for man. No one would deny that in adorning the dwelling places on earth, God was constantly thinking of their future tenants, since the purpose of the world and of all creation is man. (*Finis enim et mundi et omnis creationis homo est*). I believe that it is for this very reason that God chose the earth, designed as it is for bearing and nourishing the Creator's true image, for revolving amidst the planets, in such a way that equal numbers of them are found inside and outside its orbit.' (*Mysterium Cosmographicum*, Chapter IV.)

Dedication of the first edition of the Mysterium Cosmographicum

To their Illustrious, High-born, Noble and Righteous Lords, Sigismund Friedrich, Baron of Herberstein, Neuberg and Guttenhag; and Lankowitz, Lord Chamberlain and Lord High Steward of Corinthia, Councillor to His Imperial Majesty and to the Most Illustrious Archduke of Austria, Steward of the Province of Styria

and

to the Most Noble Lords of the Illustrious Estates of Styria, the Honourable Council of Five, my gentle and gracious Lords

Greetings and Humble Respects!

What I have promised seven months ago, to wit a work that according to the judgment of the learned will be elegant, impressive, and far superior to all annual calendars, I now present to your gracious company, my noble Lords, a work that though it be small in compass and but the fruit of my own modest efforts, yet treats of a wondrous subject. If you desire maturity—Pythagoras has already treated of it some 2000 years ago. If you desire novelty—it is the first time that this subject is being presented to all mankind by myself. If you desire scope—nothing is greater or wider than the Universe. If you desire venerability—nothing is more precious, nothing more beautiful than our magnificent temple of God. If you wish to know the mysteries—nothing in Nature is, or ever has been, more recondite. *It is but for one reason that my object will not satisfy everybody, for its usefulness will not be apparent to the thoughtless.* I am speaking of the Book of Nature, which is so highly esteemed in the Holy Scriptures. St. Paul admonished the Heathens to reflect on God within themselves as they would on the Sun in the water or in a mirror. Why then should we Christians delight the less in this reflection, seeing that it is our proper task to honour, to revere and to admire God in the true way? Our piety in this is the deeper the greater is our awareness of creation and of its grandeur. Truly, how many hymns of praise did not David, His faithful servant, sing to the Creator, who is none but God alone! In this his mind dwelled reverently on the contemplation of the Heavens. The Heavens, he sings, declare the glory of God. I will consider Thy heavens, the work of Thy hands, the moon and the stars which Thou has ordained. God is our Lord, and great is His might; He counteth the multitude of the Stars, and knoweth them by their names. Elsewhere, inspired by the Holy Ghost and full of joyousness, he exclaims to the Universe: Praise ye the Lord, praise Him, Sun and Moon, etc. Now, do the heavens or the stars have a voice? Can they praise God as men do? Nay, when we say that they themselves give praise to God, it is only because they offer men thoughts of praise to God. Thus, in what follows, let us free the very tongues of the heavens and of nature so that their voices may resound all the louder; and when we do so let no one accuse us of vain and useless efforts.

I need not stress how important a witness my subject is for the act of creation, questioned as it is by philosophers. For here we may behold how God, like a master-builder, has laid the foundation of the world according to order and law, and how He has measured all things so carefully, that we might well judge it is not nature that human art copies, but that God in His very creation was thinking of the way in which man yet unborn would be building one day.

Indeed, must we assess the value of divine things like we do a dessert, by the farthing? But, you may object, what good is an understanding of nature, what good the whole of astronomy, when the stomach is empty? However, reasoning men will not listen to the clamours of the uneducated that we cease from such studies because of this. We tolerate the painter because he delights the eye, the musician because he delights the ear, even though they bring us no other benefits. Indeed, the delight caused by their works not only benefits man, but is his glory also. What lack of education, what stupidity is it then, to begrudge the mind its own honourable joy, when we allow it to the eyes and to the ears! He who attacks these delights attacks Nature herself. For has not the all-merciful Creator, who fashioned Nature out of the Void, given every creature all it needs, and beauty and pleasure beyond in overflowing measure? Would He then single out the mind of man, the crown of all creation and made in His own image, to be alone without inspired joy? Indeed, we do not ask for what useful purpose birds do sing, for song is their pleasure since they were created for singing. Similarly we ought not to ask why the human mind troubles to fathom the secrets of the heavens. Our Creator has added mind to our senses not simply so that man might earn his daily keep—many kinds of creatures possessing unreasoning souls can do this much more skilfully —but also so that from the existence of the things which we behold with our eyes, we might delve into the causes of their being and becoming, even if this might serve no further useful purpose. And just as the bodies of men and of all other creatures are maintained by food and drink, so man's soul, which is quite different from his body, is maintained, enriched and, as it were, helped in its growth, by the food of understanding. Therefore he who is disinterested in such matters, is more alike to a corpse than he is to a living man. Now, just as Nature sees to it that the living will not lack victuals, so we may say justly that the diversity in the phenomena of Nature is so great, and the treasures hidden in the heavens so rich, precisely in order that the human mind shall never be lacking in fresh nourishment, in order that man become not satiated with the old nor stay at rest, but rather that he find the world an ever-open workshop for matching his wits.

Now what little I have served myself from the all-too-splendid table of the Creator by no means loses in value because it is despised by the great majority. More people praise the goose than do the pheasant; for the former is known to all—the latter to but a few; yet no epicure will esteem the pheasant for less than the goose. Thus the worth of my subject is needs the greater the fewer will sing its praise, if only these few be connoisseurs. What is meet for a Prince by no means suits the multitude; astronomy is not food for everybody, without distinction, but only for the aspiring soul, and this not through any fault of mine, not because of its nature, nor yet because God is a jealous God, but because most men are stupid and craven. Princes are accustomed to introducing an especially delicate dish between courses which, in order to avoid repletion, they enjoy after they are satisfied. Thus even the most noble and the wisest of men will find this and similar research to his taste only when he leaves his dwelling, and when passing through villages, cities, lands and kingdoms, he raises an enquiring eye to the great sphere of the whole earth, wishing to acquire a precise knowledge of all things. If in this he should fail to discover inspiration or lasting value in any of man's work, anything to still his hunger and to satisfy him, then will he hasten to seek out better things, then will he rise from the earth to the heavens, then will he immerse his spirit, troubled by empty worries, into that great tranquillity and proclaim with Lucretius:

> Happy the soul, whose duty it was, all this to uncover,
> Who first rose up into the heavenly heights.

He will start despising what before he deemed important; now will he esteem fully the works of God's hands, and in their contemplation he will finally come upon undisturbed and pure joy. However much and however deeply this striving may be despised, however much men may seek wealth, treasures and happiness, astronomers want nothing but the glory of knowing that their writings are for the wise and not for the rabble, for Kings and not for shepherds. Without hesitation I proclaim that there will yet be men for whom this will be a solace in their old age, men who carry out their public duties so that later they may taste without qualms of conscience the joy of which I have spoken.

Yes, there will once again come a Charles, who as ruler of Europe will seek in vain that which he, tired of ruling, finds in the narrow cell of his monastery; who among all the festivities, titles, triumphs, riches, cities, and kingdoms, finds so great a joy in the planetary sphere constructed after Pythagoras and Copernicus, that he renounces the whole world for it and prefers ruling the heavenly orbits with his measuring instruments to governing people with his sceptre. . . .

> Written on the 15th May, on which self-same day one year earlier I commenced my work.
>
> Your Highnesses' most devoted servant
> M. Joannes Keplerus of Wurttemberg
> Mathematician at Your School in Graz.

Kepler not only considered nature as the work of God but he considered it meaningless to ask questions about the material world without reference to God. Nature is understood by man's mind through quantity, and through quantity nature becomes known in its spiritual essence. In a letter to Herwart von Hohenburg on the 14th September, 1599, he wrote:

'Not every intuition is false. For man is made in God's own image and it is easily possible that in certain things which embellish the world, he is like God. For the world partakes of quantity, and the mind of man (of the world, yet greater than it) conceives nothing so well as *quantities*, for the understanding of which it has obviously been created.'

In the second chapter of the *Mysterium Cosmographicum*, which is quoted below, it is stressed that bodies can be understood through quantities, the quantitative aspect forming the starting point for a conceptual determination that allows the human mind to apprehend God's work. Thus Kepler is concerned to derive effects determined *a posteriori*—by experience ('just as a blind man feels his way with a staff')—from *a priori* reasons, *i.e.*, as stemming from the causes.

Sketch of My Principal Proof

In order now to come to my subject, and to consolidate by means of a new proof the doctrines of Copernicus about the new world which we have just discussed, I wish to run briefly through the whole subject from the beginning.

In the beginning God created the body. If this is understood then it will be reasonably clear also why God began with the creation of the body and not with anything else. Quantity, I say, lay at God's hand, and for its realization He needed everything that is of the essence of bodies, so that, as it were, the quantity of a body, inasmuch as the latter is a body at all, becomes the form and foundation of this concept. God desired quantity to come into existence before all else, so that there could be a distinction between *the curved*[1] and *the straight*. Cusanus and others seem to me so divinely great, precisely because they paid so much attention to the relationship between the straight and the curved, and because they dared to equate the curved with God, and the straight with His creations. Thus, the work of those who wish to understand the Creator through His creatures, God through men, and divine thought through human thought, is no more useful than that of those who wish to understand the curved through the straight, and the circle through the square.

Why is it that in adorning the world, God reflected on the differences between the curved and the straight, and preferred the nobility of the curved? Why, indeed? Only because the most perfect builder must needs produce a work of the greatest beauty, for it is not now, nor ever was, possible (as Cicero, following Plato's Timaeus, shows in his book on the Cosmòs) that the best should ever be anything but the most beautiful. Now, since the Creator conceived the world in His mind (we speak after the manner of men so that we mortals may understand) and since as I have said previously, the idea itself had been present beforehand and had been complete in its content, for the form of the work about to be created to become perfect also, according to the laws which God prescribes for Himself in all His goodness, it is clear that He could not derive the idea for the foundation of the world from anything but His own essence. How excellent and divine this is can be appreciated in two ways: first in itself, since God is One in His essence and Three in Person, and secondly, in comparison with His creatures.

This picture, this idea, God wished to imprint upon the world. So that the world become a most perfect and most

[1] Curved is here equivalent with circular or elliptical—*Editor*.

beautiful world, so that man could know of these ideas, the all-wise Creator created magnitude and designed quantities, whose whole essence, so to speak, lies in the distinction between the two concepts of the straight and the curved. We are thus to be made aware in the above-mentioned twofold way, that the curved represents God. Nor must we think that so purposeful a distinction in the representation of God took place as if God had not reflected on it and had created magnitude-as-body for quite different reasons and purposes, and that the distinction between straight and curved and the similarity of the latter to God came about by itself, as it were, by accident.

Rather is it probable that God from the very beginning and purposely has selected the curved and the straight for stamping the world with the divinity of the Creator; quantity had existed so that these two might be possible, and for quantity to be understood He created the body before all else.

Let us now see how the perfect Creator used these quantities in the building of the world and what, according to our lights, seems to have been His probable procedure. This then we wish to search out in hypotheses old and new, and we shall give the palm to him who will show us the way.

The fact that the whole world is circumscribed by a sphere has already been discussed exhaustively by Aristotle (in his book on the Heavens), who based his proof particularly on the special significance of the spherical surface. It is for this very reason that even now the outermost sphere of fixed stars has preserved this form, although no motion can be ascribed to it. It holds the Sun as its centre in its innermost womb, as it were. The fact that the remaining orbits are round can be seen from the circular motions of the stars. Thus we need no further proof that the curved was used for adorning the world. While, however, we have three kinds of quantity in the world, *viz.*, form, number, and content of bodies, the

THE BEGINNINGS OF MODERN SCIENCE

curved is found in form alone. In this, content is not important since one structure inscribed concentrically into a similar one (for instance, sphere into sphere, or circle into circle) either touches everywhere or not at all. The spherical, since it represents an absolutely unique quantity, can only be governed by the number Three. Thus, if, in His creation, God had been concerned with the curved alone, there would be nothing in the Cosmos except the Sun in the centre as the picture of the Father, the sphere of fixed stars (or the waters of the Mosaic story) on the surface as the picture of the Son, and the all-pervading heavenly ether, *i.e.*, extension and firmament, as the picture of the Holy Ghost. Now, since the fixed stars are innumerable while the planets have a very definite number, and since the magnitudes of the individual heavenly orbits are different, we must needs seek the cause for this in the concept of the straight. We should otherwise have to assume that God had created the world haphazard although He had the best and most reasonable plans at His disposal, and no one will be able to convince me of this even in the case of the fixed stars whose positions seem to be the most irregular, like seeds scattered at will.

Therefore let us transfer our attention to straight quantities. Just as previously we chose the spherical surface precisely because it was the most perfect quantity, so shall we now leap to those bodies which are the most perfect among straight quantities and which consist of three dimensions. It is, after all, a certain fact that the idea of the world is perfect. Thus we shall omit straight lines and surfaces, for since they are innumerable and therefore completely unsuitable for order, they are best left out of the finite, best-regulated and most perfectly beautiful of all worlds. We shall do no more than select from the infinitely many kinds of bodies some that have special characteristics; I am thinking of those whose angles, edges or side surfaces are equal either individually, in pairs, or according to some law, so that we might arrive with good reasons at some finite result. If now a class of body, defined by certain conditions, although having a finite number of types, nevertheless has a tremendous number of individual members, then whenever possible, we shall use the edges and midpoints of the faces of these bodies for the representations of the number, magnitude and position of the fixed stars. However, should this transcend our human powers we shall postpone any attempt to explain the number and position of the fixed stars until such a time as someone will be able to describe all of them without exception according to number and magnitude. Let us therefore omit the fixed stars, and leave them to the wisest of Master-builders Who alone knows the number of the stars, calling each by its name, and let us rather turn our glance towards the nearer planets which exist in much smaller numbers. If now in our final selection of bodies we omit the great mass of irregular ones, and only retain those whose faces are equal in side and in angle, there remain those five regular bodies to which the Greeks gave the following names: the Cube, or the hexahedron; the Pyramid, or the tetrahedron; the dodecahedron; the icosahedron; and the octahedron. That there can be no more than these five can be seen from Euclid, Book XIII, in the note on proposition 18.

Since the number of these bodies is well-determined and very small, while others are uncountable or infinite, there had to appear two kinds of stars which are distinguished by an obvious characteristic (such as are rest and motion). One kind must border on the infinite just as the number of fixed stars, the other must be closely limited, as is the number of the planets. Here it is not our task to find reasons why these move while those do not. However, once it is granted that planets require motion it follows that if they were to retain these, they had to be assigned curved orbits.

THE BEGINNINGS OF MODERN SCIENCE 83

Thus we come to curved orbits through motion, and to bodies through number and magnitude. We can but exclaim with Plato that *God is a great geometrician*, and in constructing the planets He inscribed bodies into circles and circles into bodies until there remained not a single body that was not endowed with movable circles internally as well as externally. From propositions 13, 14, 15, 16, and 17 of the XIIIth Book of Euclid we can see how highly suited are these bodies from their very natures for this process of inscription and circumscription. If now the five bodies be fitted into one another and if circles be described both inside and outside all of them, then we obtain precisely the number six of circles.

Now, if at any time the order of the world has been investigated on the basis of the fact that there exist six movable orbits about an immovable Sun, astronomy, at any rate, has omitted to do so. *Now, Copernicus has taken just six orbits of this kind, pairs of which are precisely related by the fact that those five bodies fit most perfectly into them, and this is the sum total of what follows.* Thus, we must heed Copernicus until someone else devises theories that will fit our philosophical conclusions even better, or until somebody will show us how that which could only be discovered by the best logic from the principles of nature could by mere chance have stolen into numbers or into the human mind. *For what could be more astonishing, what a more striking proof than the fact that what Copernicus concluded and interpreted*, a posteriori *from the phenomena, and from effects, (just like a blind man feeling his way with a staff, as he was wont to say to Rhaeticus, more through happy intuition than through reliable logical procedures) could all, I aver, have been best determined and understood with reasons which*, a priori, *stem from the causes, from the very idea of creation.*

> J. Kepler: *Mysterium Cosmographicum*, Chapter 2, pp. 45–49.

Two things will strike the modern reader who has definite ideas about contemporary science:

1. For Kepler, *science is not at all a means of procuring material benefits for man, or of making possible a technology* for better living in our imperfect world, and opening up the paths of progress. Quite the contrary —science is a means of elevating the mind, a way of finding peace and solace in the contemplation of the eternal perfection of the Creation.

2. Closely related to the above is the *remarkable disdain of empirical facts*. Experience is but the accidental discovery of relations, that can be understood much better from an insight into *a priori* reasons. The complete correspondence between 'things of the senses'—*i.e.*, the works of God—and mathematical and intelligible laws—the 'thoughts' of God— becomes the fundamental concept of the '*Harmonices mundi*'. *Platonic and neo-platonic considerations lead Kepler to the conviction that reading the works of God—in nature—is nothing but the understanding of the relationship between quantities and geometric forms.* 'Geometry, eternal like God, and shining forth from the divine Spirit, supplied God with the pictures for completing the world, so that it might be the best, the most beautiful, and the world that most closely resembled the Creator.' (*Harmonices mundi*)

THE BEGINNINGS OF MODERN SCIENCE

Bibliography:

Mysterium Cosmographicum, 1596; *Ad Vitellionem paralipomena*, 1604; *Astronomia Nova*, 1609; *Dissertatio cum nuncio sidereo*, 1610; *Dioptrice*, 1611; *Harmonices mundi*, 1619; *Epitome Astronomiae Copernicanae*, vols. 1–3: 1618, vol. 4: 1620; *Letters*, edited by Caspar and V. Dyck, Munich, 1930.

M. Caspar: *Bibliographia Kepleriana*, 1936; Carola Baumgard: *J. Kepler, life and letters*; Guenther: *Biographical Sketch*; J. Drinkwater: *Life of Kepler*; Herz: *Kepler's Astrology*; R. Small: *An account of the astronomical discoveries of Kepler*; Sir David Brewster: *Life of Kepler*.

GALILEO GALILEI
(1564–1642)

Although Galileo lived at almost the same time as Kepler, his works have quite a different air, for they plunge us directly into the beginnings of modern scientific thought.

Whenever a scientist makes a profound study of particular processes in nature, he comes to realize that *detailed processes can be isolated from total relations and that they can be recognized and formulated by means of mathematics*: in science, whose conclusions are necessary and universally valid, there is no room for human arbitrariness. Thus, we read in the *Dialogo massimi sistemi* that Nature does not create first the human spirit, and afterwards things so that these might suit it, but rather the opposite is the case. *Our assertions must always follow on observation and on experience;* in this our senses—the tools—are of primary importance. It follows that nature can only be known in particular sectors. Those, who are not prepared for such modesty of observation and description, are condemned to knowing nothing at all.

Experiment must confirm the properties of bodies, so that there is agreement between definitions and phenomena. In a letter to Carcarille of the 5th June, 1637, we can read the following: 'Now, if experiment should show that such properties as we have derived are confirmed by the free fall of natural bodies, then we can assert, without fear of error, that the concrete motion of falling is identical with that which we have predicted: if this is not the case, then our proofs lose none of their force and value, for they were designed for our assumptions alone, just as Archimedes' statements about spirals are not invalided by the fact that no natural body exists to which such spiral motion could be attributed.'

Here we can see the remarkably clear and concise expression of the *basic relationship between hypothesis and experiment*. The mind evolves assumptions for the observation of nature, which must be valid mathematically and logically. However, this validity itself is no proof of the real existence in nature of the relations implied by the assumptions. *Only when the latter are used as empirical hypotheses and are proven by experiment do they assume the character of natural laws*. Assumptions which are mathematically or logically valid, but which have no applications in nature, lose none of their validity, but do not constitute a natural law.

Even Leonardo da Vinci (1452–1519) had already rejected any thought not guided by the criterion of observation. Not observation alone, of course, for observation is only fruitful if carried out on the basis of hypotheses, but these hypotheses must in turn be confirmed by experiment. Thus, he contended that reasons (*ragioni*), *i.e.*, starting points for questions about nature, existed for every experimental discovery. Whatever is shown by experiment is thus always a limited answer of nature. Whenever reasons exist, we can give them a precise mathematical form. For Leonardo, mathematics had become the decisive link between human understanding and the reality of nature.

What is new in all this is the fact that here we are no longer concerned with observation of nature alone, but with *observation guided by certain principles* and definite rules of thought. This is nothing else but experimental observation for determining whether and to what extent certain theoretical concepts agree with observation.

Galileo distinguishes between an extensive and an intensive understanding of phenomena. By intensive understanding he refers to the fact that modern science proceeds step by step, while extensive understanding is the immediate perception of the whole from its original cause, an understanding which, in the final analysis, belongs to God alone.

a. Galileo's Self-defence Against Tradition

In order to introduce these thoughts and methods, Galileo had, above all, to counter the objections of Christian traditionalists and of the upholders of pseudo-Aristotelian science. In his famous letter to Ella Diodati, and also in many passages of his *Dialogo dei massimi sistemi*, we have evidence of his pathetic efforts to free himself from bondage to a fossilized tradition:

Florence, 15th January, 1633.

When I ask: whose work is the Sun, the Moon, the Earth, the Stars, their motions and dispositions, I shall probably be told that they are God's work. When I continue to ask whose work is Holy Scripture, I shall certainly be told that it is the work of the Holy Ghost, *i.e.*, God's work also. If now I ask if the Holy Ghost uses words which are manifest contradictions of the truth so as to satisfy the understanding of the—generally uneducated—masses, I am convinced that I shall be told, with many citations from all the sanctified writers, that this is indeed the custom of Holy Scripture, containing as it does hundreds of passages that taken literally would be nothing but heresy and blasphemy, for in them God appears as a Being full of hatred, guilt and forgetfulness. If now I ask whether God, so as to be understood by the masses, had ever altered His works, or else if Nature, unchangeable and inaccessible as it is to human desires, has always retained the same kinds of motion, forms and divisions of the Universe, I am certain to be told that the Moon has always been round, even though it was long considered to be flat. To condense all this into one phrase: Nobody will maintain that Nature has ever changed in order to make its works palatable to men. If this be the case, then I ask why it is that, in order to arrive at an understanding of the different parts of the world, we must begin with the investigation of the Words of God, rather than of His Works. Is then the Work less venerable than the Word? If someone had held it to be heresy to say that the Earth moves, and if later verification and experiment were to show us that it does indeed do so, what difficulties would the church not encounter! If, on the contrary, whenever the works and the Word cannot be made to agree, we consider Holy Scripture as secondary, no harm will befall it, for it has often been modified to suit the masses and has frequently attributed false qualities to God. Therefore I must ask why it is that we insist that whenever it speaks of the Sun or of the Earth, Holy Scripture be considered as quite infallible?

Galileo's Dialogue about the Two Chief Systems of the World[1]

The First Day:

SAGREDO: I ever accounted extraordinary madness that of those who would make human understanding the measure of what Nature has the power or knowledge to effect; whereas, on the contrary, there is no least effect in Nature which can be fully understood by the most speculative mind in the world. Their vain presumption of knowing all can spring only from their never having known anything; for if one has but once had perfect knowledge of one single thing, and but truly tasted what it is to know, one shall see that of the infinite other conclusions, one understands not so much as one.

[1] Based on the translation by T. Salusbury, Esq., 1661.

SALVIATI: Your discourse is very conclusive, and this is borne out by the example of those who do understand, or have known something, which the more knowing they are the better they know, yet freely confess that they know little; nay, the wisest man in all Greece, and for such pronounced by the Oracle, openly professed to know that he knew nothing.

SIMPLICIO: It must be granted therefore, either that Socrates or that the Oracle itself is a liar, in declaring him most wise who himself professes to be most ignorant.

SALVIATI: Neither one nor the other follows, for both may be true. The Oracle judged Socrates to be the wisest of men, whose knowledge is limited; Socrates acknowledged that he knew nothing in relation to absolute wisdom, which is infinite; but to infinity much is no more than is little or nothing—for to arrive at the infinite number, it is immaterial whether we add thousands, tens or zeros—therefore Socrates truly saw that his wisdom was nothing compared with the infinite knowledge he desired. But yet, because there is some knowledge found amongst men, and this not equally shared between all, Socrates might have a greater share than others, and thus bear out the answer of the Oracle.

SAGREDO: I think I understand this well; Simplicius, there is a power of acting but not all men have it equally; and it is obvious that the power of an emperor is far greater than that of a private person; but both are as nothing compared with Divine Omnipotence. Amongst men, there are some who understand agriculture better than others; but what has the knowledge of planting a vine in a trench to do with the art of making it sprout, of obtaining food, and selecting from it one part for the leaves, another for the sprouts, another to make grapes, another to make raisins, another to make their skins—all this is the work of most wise Nature. This is but one act of the innumerable which

Nature does, and in it alone is discovered an infinite wisdom, so that Divine Wisdom may be concluded to be infinitely infinite.

SALVIATI: Here is yet another example. Do we not say that in his clever discovery of a most lovely statue in a piece of marble, the wit of Buonarrotti has been raised high above the vulgar wits of other men? And yet such work is but the external and superficial imitation of but a single posture of one motionless man; but what is this in comparison with a man made by nature, composed of as many internal as external members, of so many muscles, tendons, nerves, bones, which serve for so many different motions? And what shall we say of the senses, and of the powers of the soul, and lastly, of the understanding? May we not say, and that with reason, that the structure of a statue falls far short of the formation of a living man, yea, even of a contemptible worm?

SAGREDO: And what difference think you, was there between the Dove of Archytas,[1] and one made by Nature?

SIMPLICIO: Either I am none of your knowing men, or else there is a manifest contradiction in your discourse. You account understanding amongst the greatest—if not *the* greatest—of Nature's gifts to man, and yet you said a little earlier, with Socrates, that he had no knowledge at all; therefore you must say Nature also did not understand how to make an understanding mind.

SALVIATI: You argue very cunningly, but to reply to your objection, I must have recourse to a philosophical distinction, and say that the understanding is of two different kinds, *i.e.*, intensive or extensive. Compared with extensive understanding, *i.e.*, the multitude of intelligibles, which are infinite, the understanding of man is as nothing, though he

---

[1] A Pythagorean philosopher and statesman, *c.* 400–365 B.C. He is alleged to have constructed a model pigeon that could fly.

understand a thousand propositions for a thousand is as nothing to infinity: but taking intensive understanding, in so far as this term implies the intensive, *i.e.*, perfect understanding of some propositions, I say that the human mind can understand some propositions so perfectly, and be so absolutely sure of them as Nature herself. Such are the pure mathematical sciences, *viz.*, Geometry and Arithmetic, in which Divine Wisdom knows infinitely more propositions because it knows them all; but I believe that the knowledge of those few comprehended by human understanding, is equal to the Divine as to their objective certainty; for it has arrived at the comprehension of the necessity, than which there can be no greater certainty.

SIMPLICIO: This seems to me a very bold and rash assertion.

SALVIATI: These are common notions, and far from all presumption or boldness, they in no manner detract from the Majesty of Divine Wisdom, for nothing diminishes Its omnipotence when we say that God cannot undo what He has once done; but I suspect, Simplicio, that your scruples arise from your having misunderstood me; therefore, the better to express myself, I say that as to the truth, of which mathematical demonstrations give us the knowledge, it is the same as known to Divine Wisdom, but this I must grant you, that the manner in which God knows infinite propositions, of which we understand a few, is much more excellent than ours, which proceeds by ratiocination, and passes from conclusion to conclusion, whereas His is done at one single thought or intuition; and whereas we, for instance, to attain the knowledge of some property of the circle, which has an infinite number of them, beginning from one of the most simple, and taking that for its definition, proceed from one argument to another, and then to a third, and then to a fourth, *etc.*, the Divine Wisdom understands by the apprehension of Its essence and without temporary ratiocination all these infinite properties. In reality, these are virtually contained in the definition of all things; perhaps, although infinite in number, they are yet in their essence but one alone in the Divine Mind. All this is not entirely alien to the human understanding, but only beclouded by thick mists, which are partly dispelled and clarified whenever we are made masters of any conclusions, firmly demonstrated and so perfectly made ours, so that we can speedily run through them. For what other is the proposition that the square of the side subtending the right angle in any triangle is equal to the squares of the other two, which include it, but that parallelograms standing on the same base and between the same parallels are equal to one another? And is this not the same as saying that two surfaces are equal whenever they can be superimposed the one upon the other without gap or overlap? Now, these inferences which our minds learn but slowly and gradually, are penetrated by the Divine Wisdom like light, in an instant; or in other words they are forever contained in It; that our understanding both as to the way and the number of things comprehended, is infinitely surpassed by Divine Wisdom. Yet I do not so despise our understanding as to consider it as absolutely naught; indeed, when I consider how many and what great mysteries men have understood, discovered and contrived, I very plainly know and understand man to be one of the most excellent works of God.

SAGREDO: I have often reflected, in this connection, how great is man's wit; and whilst I run through the great many of his admirable inventions, both in Arts and in Science, and reflect on my own mind which is far from promising me the discovery of any new thing, or indeed even the understanding of what has already been discovered, I am confounded with wonder and despair, and I account myself little less than miserable. If I behold a statue of some excellent master, I say to myself: When will I know how to chisel away the refuse of a piece of marble and discover so lovely a figure as is hidden therein? When will I mix and spread so many lovely colours upon a cloth or wall and represent therewith all visible objects like a Michelangelo, a Rafael or a Titian? If I behold what inventions men have made in comparting musical intervals and in establishing precepts and rules so that they might be an admirable delight to the ear: When shall I cease to marvel? What shall I say of the great many instruments of that art? The reading of excellent poets, with what admiration does it swell anyone who attentively follows the invention and explanation of their fancies? What shall we say of architecture? What of navigation? But above all the other stupendous inventions, what sublimity of mind had he who devised a way of communicating his most secret thoughts to another, though very far distant from him in either time or place, speaking to those who are in India, to those not yet born, nor yet to be born within a thousand or even ten thousand years? And with what ease? By the mere manipulation of some twenty signs on paper. Let this be the crown of all the admirable inventions of man and the close of our discourse for today: For the warmer hours being past I assume that Salviati wishes to take the air in his gondola; but tomorrow we shall both wait on you and continue the discourses we have begun.

The Second Day:

SALVIATI: Our diversions of yesterday have led us out of the path of our discourse so far and so often, that I know not how I can without your assistance recover the track on which I am to proceed.

SAGREDO: I do not wonder that you, whose mind is charged both with what has been and also with what is still to be spoken, do find yourself in some confusion, but I, a mere listener, who have to burden my memory with such things only as I have heard, may hope by a succinct rehearsal to recover the main thread of our discourse. As far, therefore, as my memory serves me, the sum of yesterday's conversations was the examination of the principles of Ptolemy and Copernicus, and which of their opinions is the more probable and rational: that which asserts the substance of celestial bodies to be ingenerable, incorruptible, inalterable, impassible, and, in a word, exempt from all kinds of change save that of place, and therefore to be a fifth element, quite different from our elementary bodies which are generable, corruptible, alterable, *etc.*, or else the other view, which taking away such deformities from parts of the world, holds the Earth to enjoy the same perfections as the other integral bodies of the universe; and esteems it a movable and erratic globe, no less than the Moon, Jupiter, Venus or any other planet; and lastly draws many a particular parallel between the Earth and the Moon, indeed more in the case of the Moon than any other planet, since by virtue of its greater closeness we can observe it more certainly. And having concluded this second opinion to be more probably correct than the first, I should think it best if we were to go on to examine whether the Earth is immovable, as had hitherto been believed by most men, or else movable, as some ancient philosophers held, and also others of more recent times; and if it be movable to enquire of what kind its motion may be.

SALVIATI: Now I know what part we have to pursue, but before we proceed further, I must say something to you touching upon your last words, how the opinion that the Earth is of the same condition as the celestial bodies seems to be more true than the contrary; however I have not myself affirmed any such thing nor yet the contrary position but merely wished to present on both sides those reasons, answers, arguments and solutions which have hitherto been thought out by others, together with some which I have stumbled upon in my long researches on this question, but always leaving the decision to the judgement of others.

SAGREDO: I have unawares been carried away by my own feelings in the matter; and believing that others would judge as I did, I made general a conclusion which should have remained particular; and therefore I confess my error, particularly since I do not even know Simplicio's judgement in the matter.

SIMPLICIO: I must confess that I have reflected all night upon what happened yesterday, and truth to tell, I have discovered many shrewd, new and plausible notions; yet nevertheless I find that I am swayed by the opinion of so many great writers, and in particular . . . I see you shake your head, Sagredo, and smile to yourself as if I had uttered some great absurdity.

SAGREDO: I not only smile, but to tell you the truth I am bursting with restraining myself from laughing outright; for you have reminded me of a very pretty story that I have witnessed not so many years ago, in the company of some of my other worthy friends, whom I could name to you.

SALVIATI: It would be well if you told us about it, so that Simplicio might not think it was he who caused your laughter.

SAGREDO: Very well. One day, in his home at Venice, I found a famous physician, to whom some flocked for their studies and others out of curiosity to see some dissections by the hands of this learned, careful and experienced anatomist. On that day, it so happened that, while I was there, he was in search of the origin and rise of the nerves about which there is a famous controversy between the Galenists and the Peripatetics. While the anatomist demonstrated how the great many nerves starting from the brain, their root, and then passing by the nape of the neck, distend thereafter along the backbone and branch out through all the body and that a very small filament, as fine as a thread, went to the heart; he turned to a gentleman whom he knew to be a peripatetic philosopher, and for whose sake he had, with extraordinary exactness, demonstrated and proved everything, and asked him if he was at length satisfied and persuaded that the origin of the nerves proceeded from the brain, and not from the heart. To which the philosopher, after he had stood musing awhile, answered: 'You have made me see this business so plainly and sensibly, that did not the text of Aristotle assert the contrary, which positively affirmed the nerves to proceed from the heart, I should be constrained to confess your opinion to be true.'

SIMPLICIO: I would have you know, masters, that this controversy about the origin of the nerves is by no means as proved and decided as some may perhaps persuade themselves.

SAGREDO: Nor doubtless will it ever be, if it find such contradictions; but what you say by no means lessens the extravagance of the answer of that Peripatetic, who against such sensible experience produced not other experiments nor yet the arguments of Aristotle, but his mere authority and pure *ipse dixit*.

SIMPLICIO: Aristotle has gained his authority by the force of his demonstrations and by the profoundness of his arguments; but it is essential that we understand him, and not only understand but have so great familiarity with his books that we form a perfect idea of them in our minds, so that every saying of his might always, as it were, be present in our memory; for he did not write for the crowd, nor did he feel obliged to spin out his syllogisms with the trivial method of disputes; indeed he has sometimes made free to place the proof of one proposition amongst facts which seem to treat of quite another point; and therefore it is essential to be master of all that vast idea, and to learn how to connect this passage with that and how to combine this text with another far remote from it; for it is beyond doubt that who has thus studied him knows how to gather from his books the demonstrations of every knowable deduction, for they contain everything.

SAGREDO: But, good Simplicio, just as the things scattered here and there in Aristotle give you no trouble in their collection, and in persuading yourself that by comparing and connecting several small sentences you can extract the juice of some desired conclusion, so what you and other brave philosophers do with the text of Aristotle, I could do by the verses of Virgil or Ovid, composing thereof Centones[1] wherewith I could explain the entire affairs of man, and the secrets of nature. But why do I speak of Virgil or any other poet? I have a little book, much shorter than Aristotle or Ovid, in which are contained all the sciences, and with very little study one may gather out of it a most perfect idea, and this is the Alphabet; and there is no doubt but that he who knows how to combine and dispose aright this or that vowel, with such and such consonants may gather thence the infallible answer to all doubts, and deduce from them the principles of all Sciences and all Arts; just in the same manner as the painter from diverse simple colours, laid severally upon his palette, proceeds by mixing a little of this and that with a little of a third, to represent to the life men, plants, buildings, birds,

[1] Works consisting of many fragments of verses gathered from other poets.

fishes, and in a word reproducing whatever object is visible; though there be never upon the palette either eyes, feathers, fins, leaves or stones. Indeed, it is essential that none of the things to be imitated, or any part of them, be actually among the colours if you want the latter for representing all things; for should there be, for instance, feathers amongst them, these would serve to represent nothing but birds and plumed creatures.

SALVIATI: And there are certain gentlemen yet living and in health who were present when a doctor who happened to be Professor in a famous Academy, hearing the description of the telescope, by him not seen before, said that the invention was taken from Aristotle, and causing his works to be fetched, he turned to a place where the philosopher gives the reason why, from the bottom of a very deep well, one may see the stars in Heaven at noon; and addressing himself to the company, 'See here,' he says, 'the well which represents the tube; see here the gross vapours from which is taken the invention of lenses, and see here lastly the intensification of the powers of sight by the passage of light-rays through a denser, dark and transparent medium.'

SAGREDO: This manner of understanding all that is knowable, is like that whereby a piece of marble contains in it one, indeed a thousand, very beautiful statues, but the difficulty lies in discovering them, or we may say that it is like the prophecies of the Abbot Joachim, or the answers of the Heathen Oracles, which cannot be understood till all the things foretold have come to pass.

SALVIATI: And why do you not add the predictions of the Astrologers which with equal cleverness can be seen after the event, in horoscopes or, if you will, the configuration of the Heavens.

SAGREDO: In the same way, the chemists find, with their *humor melancholicus*, that all the greatest minds in the world have, in fact, written of nothing but the way to make gold; but that they might transmit the secret to posterity without divulging it to the crowd, they contrived by one means or another how to conceal it under several masks; and it would make one merry to hear their comments on the ancient poets, discovering important mysteries hidden in their fables; and the significance of the Loves of the Moon and her descent to the Earth from *Endymion*, her displeasure with *Actæon*, and what was meant by *Jupiter's* turning himself into a shower of gold and into flames of fire; and what great secrets of art are contained in *Mercury the Interpreter*, in the thefts of *Pluto* and in the *Golden Bough*.

SIMPLICIO: I believe, and in part I know, that there is no shortage of extravagant heads in the world, whose vanities ought not redound to the prejudice of Aristotle, of whom methinks you speak sometimes with too little respect. The very antiquity and excellence of his name that he holds in the opinions of so many famous men, should suffice to render him honourable with all that profess themselves learned.

SALVIATI: You do not state the matter rightly, Simplicio. There are some of his followers who fear before they are in danger, who give us occasion, or more correctly would give us cause, to esteem him less, should we consent to applaud their caprice. And you, yourself, are you so simple as not to know that had Aristotle been present, had he himself heard the doctor who would have made him author of the telescope, he would have been more displeased with him than with those who laugh at the doctor and his comments? Do you question whether Aristotle, had he but seen the new astronomical discoveries, would not have changed his opinion, amended his books, and embraced a more sensible doctrine, rejecting those silly gulls which too scrupulously defend whatever he has said, not considering

Meonian damsels. Oh, the unheard-of baseness of servile souls!—to make themselves willing slaves to other men's opinions—to receive them for inviolable decrees—to embrace them fully satisfied and convinced by arguments that are so conclusive and efficient that they themselves are certainly unable to decide whether they were really written for a particular purpose or to prove a particular assumption, or the contrary! But, what is even madder, they are at loggerheads amongst themselves as to whether the author himself has affirmed or denied a particular proposition! What is this but to make an Oracle of a wooden idol, to run to it for answers, to fear, reverence and adore it?

SIMPLICIO: But if we should recede from Aristotle, whom are we to take for our guide in philosophy? Name me another!

SALVIATI: We need a guide when we are in unknown and uncouth lands, but in the open plains the blind alone stand in need of a leader, and for such it is better that they stay at home. But he that has eyes in his head, and in his mind, him should we choose as our guide. Yet mistake me not, thinking that because I say this I would not listen to Aristotle; on the contrary I commend the reading and diligent study of his works, and only deplore a slavish surrender to him, a blind acceptance of any of his *dicta* without any search for further reasons to take these for inviolable laws. This is an abuse that carries with it another inconvenience, to wit, that others will no longer take pains to understand the validity of his proofs. And what is more shameful in the middle of public disputes, while one person is treating of provable conclusions, than to hear the other parry with a passage from Aristotle, often written with quite a different aim, and thus to stop the mouth of his opponent? But if you will continue to study in this manner, I would have you lay aside the name of philosopher and call yourselves either historians or Doctors of Memory, for it is not fit that those who never philosophize should usurp the honourable title of philosophers. But it is best for us to return to shore, and not launch further into a boundless gulf, out of which we shall not be able to get before night. Therefore, Simplicio, come either with arguments and proof of your own, or those of Aristotle, and bring us no more texts and naked authorities, for our disputes are about the sensible world and not about paper.

b. Galileo's Conception of Modern Science

The above extracts from Galileo's writings derive their historical significance from his attitude towards tradition. His new method emerges in the following short extract from his *Dialogues and Mathematical Demonstrations on two new Sciences*. Its object is not the description of new phenomena: the motion of falling bodies had been observed from time immemorial, but the particular laws governing it had never been investigated in detail. A phenomenon is said to be governed by laws when it can be isolated from the complicated motions of bodies in general, when it can be identified precisely by means of particular measurements, principles or axioms, and when its properties can be demonstrated. *By demonstration we mean the determination and explanation of the observed phenomena with respect to a given hypothesis.* Only this constitutes true science, which is not satisfied with accidental, changeable and relative determinations. Thus, the definition of the phenomenon must correspond with the 'behaviour' of nature within the framework of the hypothesis; in other words, nature is here a *modest, sharply delimited sector of, and extract from, the multiplicity of phenomena* observed by our senses, a sector in which—as Galileo states—we allow 'nature to lead us by the hand'. Questions and answers, observations and determinations, are no longer directed at a general, metaphysical and theological understanding, but are delimited with modesty. While Kepler attributed to phenomena an eternal, metaphysical and theological character, independent of observation, in Galileo there had taken place a complete reversal of this attitude. For Kepler science was still quite unhistorical, *with Galileo it has become historical, because the properties of the phenomena to be determined are only investigated from the viewpoint of man-made hypotheses*. Whenever these hypotheses change, then there must be a corresponding change in the description of the phenomena considered within these limits. True, within particular limits set by man nature will always give the same answer, but it is just this 'eternal', immutable behaviour according to law that has now become the object of scientific striving, and its realization is the pride of the scientist.

Galileo's Dialogues of Motion

The Third Day:

We have developed a completely new science about a very old subject. There is nothing in nature more ancient than motion, of which many and great volumes have been written by philosophers. But yet there are sundry symptoms and properties in it worthy of our notice, which I find not to have been hitherto observed, much less demonstrated by any. Some slight particulars have been noted: as that the natural motion of heavy bodies is continually accelerating as they descend towards their centre, but it has not as yet been stated in what proportion this acceleration is produced. For no man that I know has ever demonstrated that there is the same proportion between the spaces through which a body moves in equal time, as there is between the odd members that follow the unit in sequence. It has been observed that project(ile)s (or things thrown or darted with violence) make a line that is somewhat curved; but that this line is a parabola none have hinted; yet these and sundry other things, no less worthy of our notice, will I here demonstrate: and what is more, I will open the way to a most extensive and excellent science, of which these our labours shall be the elements into which more subtle and piercing wits than mine will be better able to dive.

We divide this Treatise into three parts. In the first part we consider such things as concern uniform motion. In the second we write of motion naturally accelerated. In the third we treat of violent motion or of projectiles.

Of Motion Naturally Accelerate

In the former book we have considered the properties which accompany uniform motion, we are now to treat of another kind of motion which we call accelerate. At first it will be expedient to find out and explain a definition best agreeing with that Nature makes use of. For though it be not inconvenient to feign a motion at pleasure, and then to consider its properties (as those have done, who having formed in their imagination helixes and conchoids, which are lines arising from certain motion, although not used by Nature, and upon that supposition have laudably demonstrated the symptoms thereof), yet because Nature makes use of a certain kind of acceleration in the descent of heavy bodies, we are resolved to search out and contemplate the passions thereof, and see whether the definition that we are about to produce of this our accelerate motion, does aptly and congruously agree with the essence of motion naturally accelerate.

After many long and laborious studies we have found out a definition which seems to express the true nature of this accelerate motion, inasmuch as all the natural experiments that fall under the observation of our senses, agree with those its properties that we intend anon to demonstrate. In this disquisition we have been assisted, and as it were led by the hand, by that observation of the usual method and common procedure of Nature herself in her other operations, wherein she constantly makes use of the first, simplest and easiest of all means: for I believe that no man can think that swimming or flying can be performed in a more simple or easy way than that which fishes and birds do use out of natural instinct. Why therefore shall not I be persuaded that, when I see a stone continually acquiring new additions of velocity in its descent from rest out of some high place, this increase is produced in the simplest, easiest and most obvious manner that we can imagine? Now if we seriously examine all the ways that can be devised, we shall find no increases, no acquisitions less intricate or more intelligible than that which ever increases or makes its additions in the same manner. This happens because of the great affinity between time and motion. For as the uniformity of motion is defined and expressed by the uniformity of time and space (for we call that motion or influence uniform by which equal spaces are past in equal times) so by the same uniformity of the parts of time we may perceive that the increase of celerity in the natural motion of heavy bodies is made after a simple and plain manner; conceiving in our mind that their motion is continually accelerated uniformly and at the same rate, whilst equal additions of celerity are conferred upon them in all equal times, so that taking any equal particles of time, beginning from the first instant in which the movable body departs from rest and enters upon its descent, the degree of velocity acquired in the first and second particles of time, is double the degree of velocity that the body acquired in the first particle; and the degree of velocity that it acquires in three particles is triple, and that in four quadruple to the same degree the first time. As, for our better understanding, if a movable body should continue its motion according to the degree or moment of velocity acquired in the first particle of time, and should extend its course equably with that same degree; this motion would be twice as slow as that which it would obtain according to the degree of velocity acquired in two particles of time; so that it will not be improper if we understand the intention of the velocity, to proceed according to the extension of the time. From whence we may frame this definition of the motion of which we are about to treat:

Motion accelerate in equal or uniform proportion, I call that which departing from rest, adds on equal moments of velocity in equal times.

Of the Motion of Projects

The Fourth Day:

What are the properties of uniform motion, and also of naturally accelerated motion along any kinds of inclined planes, we have considered above. In this contemplation which we are now entering upon, I will attempt to declare, and with solid demonstrations to establish, some of the principal symptoms, and those worthy of knowledge, which befall a movable body whilst it is moved with a motion compounded of a twofold influence; to wit, of the uniform and the naturally accelerated: and this is that motion which we call the motion of project(ile)s, whose generation I constitute to be in this manner:

I fancy in my mind a certain movable body projected or thrown along an horizontal plane without any impediment: now it is manifest by what we have elsewhere said at length, that the motion will be uniform and perpetual along the said plane, if the plane be extended to infinity: but if we suppose it finite and placed on high, the movable which I conceive to be endowed with gravity, having reached the end of the plane, in proceeding further it adds to the uniform and indelible first influence that propension downwards which it receives from its gravity; and from thence results a certain motion compounded of the uniform horizontal and of the descending naturally accelerated, which I call projection. Some of whose properties we will now demonstrate. . . .

Bibliography:

Le Opere di Galileo Galilei, Edizione Nazionale, 20 vols., Florence, 1890–1900; *The Systems of the World*, translated by T. Salusbury, 1661; *Mathematical discourses concerning two sciences*, translated by Grew and Salvio, 1914; *Sidereal Messenger*, translated by E. S. Carlos, 1633.

A. Carli and A. Favoro: *Bibliografia Galileiana*, Rome, 1896; K. von Gebler: *Galileo Galilei und die Römische Kurie*, 2 vols., 1876–1877; L. Olschki: *Galileo Galilei und seine Zeit*, 1927; E. J. Dijksterhuis: *De Mechanisering van het Wereldbeld*, Amsterdam, 1950; P. Aubanel: *Le génie sous la tiare . . . Urbain VIII et Galileo*, 1929; F. Sherwood Taylor: *Galileo and the Freedom of Thought*, London, 1938; A. Koyre: *Etudés Galiléennes*, 3 papers, 1937; A. Koyre: *Galileo and Plato, Journal of History of Ideas*, vol. 4, 1943; Mary Allan Olnay: *The Private Life of Galileo*, 1869.

ISAAC NEWTON
(1643–1727)

The methodical approach, as we have traced it in Galileo's case, had now become common property. After Bacon (1561–1626) had emphasized the importance of the empirical method, the scientific observation of nature led to an increasing number of discoveries and achievements. To recall but a few of the practical results of this new approach, in 1628 William Harvey (1578–1658) discovered the circulation of the blood, in 1600 William Gilbert (1540–1603) gave the first account of magnetic phenomena in his *De magnete*, in 1643 Galilei's pupil Evangelista Torricelli (1608–1647) invented the barometer and in 1662 Robert Boyle (1627–1691) and Edme Mariotte (1620–1684) discovered the law of atmospheric pressure.

The basic causes of the phenomenon of the motion of bodies were still unknown, but the forces could be determined and calculated from laws and relations.

Hitherto men had made scientific hypotheses apparently without any reference to natural facts, but merely to their mathematical or logical validity, and subsequently made them the basis of observation. Now it had become clear that the hypotheses must no longer remain the arbitrary creations of the human mind, but that they must be carried out in close connection with the observation of nature. A scientist's genius lies precisely in the fact that in his hypotheses he recognizes simple relations between specific natural phenomena which can be generalized as mathematical concepts, and so lead to an explanation of other natural phenomena. Phenomena themselves must inspire the scientist to make the hypotheses on which to base observation and experiment.

Newton—whose concept of nature contained a decisively new factor: he not only freed nature from an all-embracing God-relatedness, but also from its narrow connection with man—at first emerged as a mere empiricist rejecting hypotheses: understanding is derived from phenomena and is generalized inductively. Perhaps Roger Cotes, publisher of the second edition (1713) of Newton's *Mathematical Principles of Natural Philosophy*, expresses Newton's approach most clearly, when he states that those who investigate physics can be divided into three classes. The first (the scholastic philosophers) endow specific things with specific and hidden properties. The second group maintains that matter in general is homogeneous, and that the particular factors differentiating individual bodies arise from very simple and easily distinguishable relations between the particles of which the body is constituted. Since, however, this group makes free to assume the existence of arbitrary forms and magnitudes in these particles and also of indeterminate positions and motions, they fall into reveries. Those who base their speculations on hypotheses, even when they proceed strictly according to the laws of mechanics, will produce nothing but fairy-tales, albeit elegant and beautiful ones. This method of investigation cannot lead to unobjectionable conclusions. Finally—and here Newton enters—there remains a third group of investigators, those who take their stand in experimental physics. True, this group wishes to derive the causes of all things from the simplest possible principles, but it *never recognizes as a principle what has never been manifest*. Thus it adopts a double method: that of analysis and synthesis. The forces of nature and the simple laws governing them are derived from a few chosen phenomena by means of analysis, and by means of synthesis they are then made to explain the properties of the remaining phenomena. This kind of investigation is precisely the one Newton chose.

Cotes then cites the famous example of the universal law of gravitation. Some had thought that the force of gravity was inherent in all bodies, others had suspected it; but Newton was the first and only one to demonstrate its existence by means of the phenomena themselves, and to give it a firm foundation through his great theory. Thus Cotes stresses the fact that true investigation means first to derive the nature of things from real causes, and then to look for the laws. Laws must not be derived from uncertain assumptions, but from observation and experiment. Whoever believes that he can discover the principles of nature or the laws of things by relying on the powers of his mind alone, or on the inner light of his reason, must either assume that the world arose by necessity and that all laws follow from this necessity; or else he must believe that, although God created the order of nature, he, a mere mortal, can see what it is best to do. *A healthy and true science*, Cotes emphasizes, *is based on the phenomena themselves, and leads us, even against our will, to such principles, that we can clearly see in them the best reflection and the ultimate sovereignty of the wisest and mightiest of Beings.*

From the phenomena themselves, and by inductive generalizations, Newton came to recognize impenetrability, motion, the momentum of bodies, the laws of motion and gravitation. The force of gravity exists and acts according to the laws he had postulated. By making postulates which allow the determination of concepts such as mass, cause, force, inertia, space, time and motion, Newton had become the first systematizer of modern science. Thus in the Preface to the first edition (1687) of his *Philosophiae naturalis principia mathematica* we read the following:

'All the difficulty of (natural) philosophy seems to consist in this—from the phenomena of motions to investigate the forces of nature, and then from these forces to demonstrate the other phenomena; and to this end the general propositions in the first and second books are directed. In the third book we give an example of this by explaining the System of the World, for by the propositions mathematically demonstrated in the former books, in the third we derive from the celestial phenomena the forces of gravity with which bodies tend to the Sun and the several planets. Then from these forces, by other propositions which are also mathematical, we deduce the motions of the planets, the comets, the moon and the sea. I wish we could derive the rest of the phenomena of nature by the same kind of reasoning from mechanical principles; for I am induced by many reasons to suspect that they may all depend upon certain forces, by which the particles of bodies, by some forces hitherto unknown, are either mutually impelled towards one another; and cohere in regular figures, or are repelled and recede from one another. These forces being unknown, philosophers have hitherto attempted to search nature in vain; but I hope the principles here laid down will afford some light either to this or some truer method of philosophy.'

Newton had created a celestial physics, without arbitrariness or miracles, self-consistent and self-sufficient, and without any sliding into the paths of materialism. He held fast to a belief in a personal God, mechanics in nature being but a means to His ends. Even when the 'great ocean' of reality remains undiscovered, the individual facts are yet fused into a great whole by their relations. Hence Newton's famous declaration: 'I do not know what I may appear to the world; but to myself I seem to have been only like a boy playing on the sea-shore, and diverting myself in now and then finding a smoother pebble or a prettier shell than ordinary, while the great ocean of truth lay all undiscovered before me.'

Philosophiae naturalis principia mathematica
(*Book III*)

Rules of Reasoning in Philosophy

Rule I. We are to admit no more causes of natural things than such as are both true and sufficient to explain their appearances.

Philosophers assert that nature does nothing in vain, and more is in vain whenever less will serve. Nature is pleased with simplicity and affects not the pomp of superfluous causes.

Rule II. Therefore, whenever possible, we must ascribe the same natural effects to the same causes.

This applies to respiration in man and beast, the fall of stones in Europe and America, the light of our kitchen fire and the sun, the reflection of light by the earth and the planets.

Rule III. The qualities of bodies which admit neither intension nor remission of degrees, and which are found to belong to all bodies within the reach of our experiments, are to be esteemed the universal qualities of all bodies whatsoever.

For since the qualities of bodies are known to us by experiments only, we must hold for universal all such as universally agree with experiments, and such as are not liable to diminution can never be quite taken away. *We are certainly not to relinquish the evidence of experiments for the sake of dreams and vain fictions of our own making, nor are we to recede from the analogy of Nature which is always simple, and consonant with itself.* We know the extension of bodies in no other way than that of our senses, nor do these reach it in all bodies; but because we perceive extension in all that are sensible, therefore we ascribe it universally to all others also. That many bodies are hard, we learn from experiment; and since the hardness of the whole arises from the hardness of these parts, we therefore justly infer the hardness of the undivided particles not only of the bodies we feel but of all others also. That all bodies are impenetrable we gather not from reason, but from sensation. The bodies which we handle we find impenetrable, and thence conclude impenetrability to be an universal property of all bodies whatsoever. That all bodies are movable, and endowed with certain powers (which we call the force of inertia) of persevering in their motions, or in their rest, we only infer from the like properties observed in the bodies which we have seen. The extension, hardness, impenetrability, mobility and force of inertia of the whole, result from the extension, hardness, impenetrability, mobility and force of inertia of the parts; and thence we conclude the least particles of all bodies to be also all extended, and hard and impenetrable and movable and endowed with their forces of inertia. And this is the foundation of natural philosophy. Moreover, that the divided but contiguous particles of bodies may be separated from one another, is matter of observation; and in the particles that remain undivided, our minds can distinguish yet lesser parts, as is demonstrated by mathematics. But whether the parts so distinguished, and not yet divided, may, by the powers of Nature, be actually divided or separated from one another, we cannot certainly determine. Yet had we the proof of but one experiment that any undivided particle, in breaking a hard and solid body, suffered a division, we might by virtue of this rule conclude that the undivided as well as the divided particles may be divided and actually separated to infinity.

Lastly, if it universally appears, by experiment and astronomical observations, that all bodies about the earth gravitate towards the earth, and they do so in proportion to the quantity of matter which they severally contain; that the moon likewise, according to the quantity of its matter, gravitates towards the earth; that, on the other hand, our sea gravitates towards the moon; and all the planets mutually one towards another; and the comets in like manner towards the sun; we must, in consequence of this rule, universally allow that all bodies whatsoever are endowed with a principle of mutual gravitation. For the argument from the appearances concludes with more force for the universal gravitation of all bodies than for their impenetrability, of which, among those in the celestial regions, we have no experiments, nor any manner of observation. Not that I affirm gravity to be essential to bodies: by their inherent forces I mean nothing but their inertial forces. These are immutable whereas gravity is diminished as they recede from the earth.

Rule IV. In experimental physics we are to look upon propositions collected by general induction from phenomena as accurately or very nearly true, notwithstanding any contrary hypotheses that may be imagined, till such time as other phenomena occur, by which they may either be made more accurate, or liable to exceptions.

This rule we must follow, so that the argument of induction may not be evaded by hypotheses.

Philosophiae naturalis principia mathematica
(End of Book III, Section V)

Of Comets

Thus much concerning God; to discourse of whom from the appearance of things, does certainly belong to natural philosophy.

Hitherto we have explained the phenomena of the heavens and of our sea by the power of gravity, but we have not yet assigned the cause of this power. This is certain, that it must proceed from a cause that penetrates to the very centres of the sun and planets, without suffering the least diminution of its force; that operates not according to the quantity of the surfaces of the particles upon which it acts (as mechanical causes usually do), but according to the quantity of the solid matter which they contain, and propagates its virtue on all sides to immense distances, decreasing always in the duplicate proportion [*i.e.*, with the square of] the distances. Gravitation towards the sun is made up out of the gravitations towards the several particles of which the body of the sun is composed, and in receding from the sun decreases accurately in the duplicate proportion of the distances as far as the orb of Saturn; as evidently appears from the quiescence of the aphelions of the planets; nay, even to the remotest aphelions of the comets, if those aphelions are also quiescent. But hitherto I have not been able to discover the cause of those properties of gravity from phenomena, and I frame no hypotheses; for whatever is not deduced from the phenomena is to be called an hypothesis, and hypothesis, whether metaphysical, or physical, whether of occult qualities or mechanical, have no place in experimental philosophy. In this philosophy particular propositions are inferred from the phenomena, and afterwards rendered general by induction. Thus it was that the impenetrability, the mobility, and the impulsive force of bodies, and the laws of motion and of gravitation, were discovered. And to us it is enough that gravity does really exist, and acts according to the laws which we have explained, and abundantly serves to account for all the motions of the celestial bodies, and of our sea.

And now we might add something concerning a certain most subtle Spirit, which pervades and lies hid in all gross bodies; by the force and action of which Spirit the particles of bodies mutually attract one another at near distances, and cohere, if contiguous; and electric bodies operate to greater distances, as well repelling as attracting the neighbouring corpuscles; and light is emitted, reflected, refracted, inflected, and heats bodies; and all sensation is excited, and the members of animal bodies move at the command of the will, namely by the vibrations of this Spirit, mutually propagated along the solid filaments of the nerves, from the outward organs of sense to the brain, and from the brain into the muscles. But these are things that cannot be explained in few words, nor are we nourished with that sufficiency of experiments which is required to an accurate determination and demonstration of the laws by which this electric and elastic Spirit operates.

Bibliography:

Philosophiae naturalis principia mathematica, 1687, Motte's translation, 1729; *Optics*, 1704; *Arithmetica universalis*, 1707; *Analysis*, 1711; *Opuscula mathematica, philosophica et philologica*, edited by J. Castillioneus, vols. 1–3, Lausanne, 1744; *Collected Works*, edited by Samuel Horsley, 5 vols., 1779–1785.

G. J. Gray: *Bibliography of the Works of Newton*, 2nd edition, 1907; F. Dessauer: *Weltfahrt der Erkenntnis (The Life and Work of Isaac Newton)*, 1945; E. N. da C. Andrade: *Sir Isaac Newton*, 1954; Thayer & Randall: *Newton's Philosophy of Nature*, 1953.

5

THE ORIGINS OF THE MECHANISTIC AND MATERIALIST WORLD-VIEW

The Application of Newton's Methods

Christian Huygens
(1629–1695)

The method of Newtonian mechanics was being applied to ever wider domains of nature. Attempts were made to isolate the details of natural processes, and to determine their 'laws'. Thus we can read in the preface to Huygens's *Traité de la lumière* (Treatise on Light—1690): '(In this work) will be found proofs of a kind that do not assure certainty in the way that geometrical proofs do and which are, in fact, very different from the latter, for here principles are verified by the conclusions derived from them, while geometricians prove their propositions by means of certain and indisputable axioms: this is due to the nature of the subjects under consideration. In a proof of this kind it is quite possible to reach so high a degree of probability that it is by no means inferior to a rigid proof. This is the case whenever *the consequences derived from postulating the principles agree fully with the phenomena known from experience*, particularly when these are numerous, and especially if we can think up and predict new phenomena that follow from the assumptions we have made, and if subsequently we find that a successful result crowns our expectations. If now all the probability-proofs agree in the case of those subjects that I have chosen to treat, as in my opinion they in fact do, then this circumstance must perforce confirm my method of research to a high degree, and it is hardly possible that things are far from what I represent them to be.'

Huygens then explains light as the motion of a particular kind of matter; in other words, he reduces its effects to mechanical reasons.

Let us stress the fact that whenever Huygens speaks of 'philosophy' he uses the word in its original sense of 'love of knowledge'. As an example of the fact that Newtonian mechanics were applied to ever wider natural phenomena we shall now quote a chapter of Huygens' *Treatise on Light*.

On the Rectilinear Propagation of Light Rays

The procedures of proof in optics, just as in all other sciences in which geometry is applied to matter, are based on truths derived from experience; *e.g.*, the fact that light rays are propagated in straight lines, that the angle of reflection is equal to the angle of incidence, and that refraction obeys the

MECHANISTIC-MATERIALIST WORLD-VIEW 123

sine rule, so well-known today and no less certain than the others.

The majority of those who have written on the different parts of optics, have been content to take these truths for granted. Some of the more enquiring strove to discover their origins and causes, seeing that they considered them as inherently wonderful effects of nature. Since, however, the opinions offered, although ingenious, are not such that more intelligent people would need no further explanations of a more satisfying nature, I wish here to present my thoughts on the subject so that, to the best of my ability, I might contribute to a solution of that part of science which, not without reason, is considered to be one of the most difficult. I acknowledge my great indebtedness to those who were the first to start dispelling the strange gloom surrounding these things, and who aroused the hope that they might yet be explained rationally. But, on the other hand, I am not a little surprised to find that very often they considered as most certain and proven, conclusions that were only too flimsy; for to my certain knowledge no one has as yet offered a satisfactory explanation of even the first and most important phenomenon of light, *viz.*, why it is propagated precisely in straight lines, and how light rays arriving from infinitely varied directions cross without impeding one another.

In this book I shall therefore attempt, according to the principles held in contemporary philosophy, to give clearer and more probable reasons for the properties, first of the rectilinear propagation of light, and second of the reflection of light when it encounters other bodies. Then I shall explain those phenomena of the rays, which in traversing different kinds of transparent bodies undergo so-called refraction; and in this I shall also treat of the effects of refraction in the air arising out of differences in the density of the atmosphere.

I shall continue by investigating the strange refraction of

light of a particular crystal brought from Iceland. Finally I shall treat of the different forms of transparent and reflecting bodies, by means of which the rays are either made to converge on one point, or else are deviated in most different ways. In this it will be seen with what ease our new theory will lead to the discovery not only of ellipses, hyperbolae and other curves, which Descartes had ingeniously suggested for this effect, but also of those figures which form one surface of a glass, when the other surface is known to be spherical, plane, or of any other shape.

There will remain no doubt that light consists of the motion of a particular kind of matter. For if we look at its origin, we shall find that here on earth fire and flame generally produce what is doubtless contained by fast-moving bodies, seeing that they too dissolve and melt numerous other very solid bodies; and if we consider the effects of converging light gathered, for instance, by a concave mirror, we shall observe that it has the same powers of heating as has fire, *i.e.*, of separating the parts of bodies; this surely points to motion, at least in true philosophy in which the causes of all natural effects are looked for in mechanical reasons. This, in my opinion, must either be done, or else we must relinquish all hope of ever understanding anything in physics.

Since, now, according to this philosophy, it is held as certain that the sense of vision is only stimulated by the impression of a certain motion of a material acting on the nerves at the back of our eyes, this is a further reason for believing that light consists of a motion of the matter between us and the luminous body. If, furthermore, we pay attention to, and weigh up, the extraordinary speed with which light spreads in all directions, and also the fact that coming, as it does, from quite different, indeed from opposite, directions, the rays interpenetrate without impeding one another, then we may well understand that whenever we see a luminous object, this cannot be due to the transmission of matter which reaches us from the object, as for instance a projectile or an arrow flies through the air, for this is too great a contradiction of the two properties of light, and the second in particular. Thus it must spread in a different way, and precisely our knowledge of the propagation of sound in air can lead us to an understanding of this way.

We know, that by means of air, which is an invisible and impalpable body, sound spreads through the whole of space surrounding its source by a motion which advances gradually from one air particle to the next, and since the propagation of this motion takes place with equal speed in all directions, spherical surfaces must be formed that spread out further and further, finally to reach our ears. Now it is beyond doubt that light also reaches us from luminous bodies by means of some motion which is imparted to the intermediate matter, for we have already seen that this could not have happened by means of the translation of a body that might have reached us from there. If now, as we shall soon investigate, light needs time for its path, it follows that this motion imparted to matter must be gradual, and that, like sound, it must spread in spherical surfaces or waves; I call them waves because of their similarity to those which we see being formed in the water when a stone is thrown into it, and because they enable us to observe a like gradual spreading-out in circles, although they are due to a different cause and only form in a plane surface.

Christian Huygens: Traité de la lumière, 1690.

Bibliography:

Oeuvres complètes, published by the Societé Hollandaise des Sciences, 22 vols., 1888–1950; *Traité de la lumière*, 1690; *Horologium oscillatorium*, 1673; *Cosmotheoros*, 1698; *Sur la Cause de la pesanteur*, 1690.

A. B. Bele: *Christian Huygens and the Development of Science*, 1948; E. J. Dijksterhuis: *Christian Huygens*, Haarlem, 1951.

GASSENDI—BOYLE—DESCARTES

With the renewed flourishing of the sciences in the seventeenth century we had the formation of learned Societies (Acadèmie Française in Paris, 1635, and the Royal Society in London, 1663). Along with the results of scientific research new trends in philosophy began to emerge. To enter into these is not our task, but we shall draw attention to three natural philosophers: Petrus Gassendi (1592–1655), Robert Boyle (1627–1691) and René Descartes (1596–1650). Their thought was to have metaphysical effects in the heyday of the mechanistic approach. Gassendi, originally a teacher of rhetoric and a professor of philosophy, taught mathematics in Paris. He thought that the atomism of Epicurus was capable of supplying a causal and mechanical explanation of natural processes. Although matter can be divided infinitely by means of mathematics, in practice we shall always arrive at indivisible atoms having hardness and impenetrability. All phenomena, their origins and their disappearances, are due to the combination and separation of these atoms, which have an inherent tendency to motion. It is important to recall that for Gassendi this atomistic picture rested on God.

Owing to Gassendi's influence, Robert Boyle (1627–1691) also accepted an atomistic explanation of nature. There exists only one kind of matter having extensibility, impenetrability and divisibility; by means of motion there are produced minute particles of a certain size, form and position, and it is these that combine to form compound bodies. Boyle, too, saw the cause of motion in God. In his *Chymista scepticus* (1668) we can read: 'If we attribute to the particles, of which every element consists, a definite magnitude and form, then it can easily be shown that these different forms of particles can be mixed in so many different proportions, and combined in so many different ways, that an almost incredible number of diversified solid bodies can be compounded from them, particularly since the particles of any one single element, by merely combining with one another, can form small masses that in magnitude and form differ from the particles of which they consist.'

For René Descartes, the road to the investigation of truth was mathematics. Starting from psychophysical dualism, *i.e.*, from the distinction between thinking substances and those having extension only, he was the first to attempt the development of the mechanics not only of the celestial vault, but also of the soul, of organic as well as inorganic nature: for him physiology was no less a *mechanical science* than was astronomy. Nature can only be explained by itself, and its laws are identical with those of mechanics. In him we can sense the growing influence of science, since he uses the results of scientific research for confirming the principles of his philosophy. Thus the urge to arrive at philosophical conclusions from scientific knowledge was becoming of increasing importance. *The modesty which granted validity to natural laws only within the framework of particular problems, or within strictly limited domains* (see p. 104), *is here renounced.*

Thus mechanistic thought gave the impetus to the gradual spread of a materialist philosophy, which was to attain its full impact during the Enlightenment, and to find its English exponent in John Locke (1632–1704). The French Enlightenment, of which the fundamental ideas were contained in the famous *Encyclopédie ou Dictionnaire raisoné des sciences, arts et métiers* (1751), was given its particular mould by Voltaire and D'Alembert. The extract on page 130 from the '*Discours Préliminaire*' of the *Encyclopedia* shows clearly that the careful attitude of classical natural philosophy, which consistently limited the domain of validity of the propositions it derived from observations, has here been discarded. What is attempted here is to show that all our ideas are reducible to sensations. Thus science has given birth to its own, *inherently uncritical, philosophy*. This will be illustrated by the utterances of some further materialist writers.

Jean Lerond d'Alembert
(1717–1783)

Matter and motion are the essential presuppositions of statics and mechanics. Scientists were proud to be able to point out that the known laws of equilibrium and motion could really be observed in the bodies surrounding us, and that therefore they must necessarily be true. All metaphysical explanations were discarded, and this is illustrated by d'Alembert in his preface to the *Traité de dynamique* (Paris 1743):

'From all these considerations it is clear that the laws of statics and mechanics expounded in this book, are the very laws which follow from the existence of matter and motion. Now, experience shows us that these laws can really be observed in the case of the bodies that surround us. The laws of equilibrium and motion, as experience has taught us, are thus of a necessary validity. A metaphysician might perhaps be satisfied with the assertion that it was due to the wisdom of the Creator, and to the simplicity of His concepts, that no other laws of equilibrium and motion were postulated than those which follow from the very existence of bodies and from their common impenetrability; however, we believe that we must *refrain from this kind of consideration*, since it would lead us to a vague principle; the nature of the Highest Being *is much too hidden from us* for us to be able to recognize what agrees with the basic principles of His wisdom and what does not; we can only suspect the effects of this wisdom through the *observation of nature*, when mathematical considerations have shown us the simplicity of these laws, and when experience has taught us the scope and domain of their validity.'

The materialist world-view, based on the laws of mechanics, had now reached maturity; nature has become a system of measurable motions and energies.

Preface to the French Encyclopaedia

As all our direct knowledge comes in by the Senses, all our Ideas are consequently due to our Sensations.

In this Study of Nature, prosecuted partly from Necessity, and partly for Amusement, we find that Bodies have numerous Properties; but generally so united in one and the same Subject, that to study each of them thoroughly, we are obliged to consider them separate. And by this Operation of the Mind, we soon discover Properties apparently belonging to all Bodies; as the Power of moving or resting, and the Communication of Motion: which are the Sources of the principal Changes we observe in Nature. Upon examining these Properties by the Senses, especially the Communication of Motion, we soon discover another Property, whereon they depend, namely, Impenetrability; or that kind of Force, by which each Body excluded all others from the Place it possesses: so that two Bodies being brought as close as possible together, can never occupy a less Space than that which each separately fills.

Impenetrability is the principal Property by which we distinguish Bodies from the Parts of indefinite Space, wherein we suppose them placed; at least, our Senses oblige us to judge thus of them: and if our Senses here deceive us, it is so metaphysical an Error, that our Existence and Preservation have nothing to fear from it; whilst, by our common manner of conceiving, we constantly return to this Notion, in spite of ourselves.

Everything induces us to look upon Space, if not as the real, yet as the supposed Place of the Bodies; and it is, actually by the Means of the Parts of Space, consider'd as penetrable, and immovable, that we come at the precisest Idea we can have of Motion: and are therefore naturally impell'd to distinguish, at least mentally, two kinds of Extension: the one impenetrable; the other constituting the Place of Bodies. And thus, though Impenetrability necessarily enters the Idea we have of the Parts of Matter; yet this being a relative Property, the Idea of which we only acquire by examining two Bodies together; we soon accustom ourselves to consider it as distinguish'd from Extension; which we can mentally abstract from Body.

In this new Consideration, we regard Bodies only as figur'd and extended Parts of Space: which is the most general, and abstract Light they can be consider'd in: for Extent wherein we distinguish no figur'd Parts, is but like a Picture out of Sight; where the whole escapes us, because we can distinguish nothing. Colour and Figure, Properties constantly, though differently, residing in Bodies, help us to distinguish Body from Space. Nay, one of these Properties is sufficient for the Purpose; insomuch that to consider Body in its most mental Form, we prefer Figure to Colour; whether because Figure is more familiar to us, as being distinguishable both by the Sight and Touch; or because it is more easy to consider the Figure of a Body without its Colour, than the Colour without Figure; or lastly, because Figure serves more easily, and more precisely, to determine the Parts of Space.

Hence we are brought to fix the Properties of simple

figur'd Extent, which constitutes the Object of Geometry; a Science that to obtain its End the better, first considers Extension limited by one, then by two, and at last by three Dimensions. These three Dimensions constitute the Essence of all mental Bodies, or Portions of Space, terminated every way by intellectual Boundaries. And thus, by the successive Operations and Abstractions of the Mind, we divest Matter of almost all its sensible Properties; in order to examine only its Idea. The Discoveries to which this Enquiry leads, cannot fail of being extremely useful; especially as it is here unnecessary to regard the Impenetrability of Bodies: for when we contemplate their Motion, we need only consider them as figur'd and movable Parts of Space, separate from each other.

As the Examination of figur'd Extension presents numerous possible combinations; it is necessary to find a Means of rendering these Combinations facile. And as they principally consist in the Calculation, and Relation, of the different Parts of which we imagine Geometrical Figures to be form'd; this Enquiry presently leads us to Arithmetic, or the Science of Numbers; which is no more than the Art of finding a short method of expressing a single relation, arising from the Comparison of many: and the different ways of comparing these Relations, give us the different Rules of Arithmetic.

By reflecting upon these Rules, we easily deduce certain Principles, or general Properties, of Relations; and by expressing these Relations in a general manner, we discover different possible Combinations; the Results whereof, being reduced to one general Form, are no more than arithmetical Combinations, indicated and represented by the most simple and concise Expression, consistent with their Generality. The Art, or Science, of thus denoting these Relations, is what we call *Algebra*. And although, properly speaking, no Calculation can possibly be made, but by Numbers; nor any Magnitude measured but Extension (for without Space we could not exactly measure Time) we arrive, by constantly generalizing our Ideas, at that principal Branch of Mathematics, and all the natural Sciences, called the *Science of Magnitude* in general; which is the Foundation of all the possible Discoveries of Quantity, or whatever is susceptible of Increase or Diminution.

This Science is the most remote End to which the Contemplation of the Properties of Matter can carry us; it being impossible to go further, without entirely quitting the material Universe. But such is the Progress of the Mind in its Enquiries, that after having generaliz'd its Perceptions, so as not to be able to analyse them further; it returns back by the same Steps, recomposes its Perceptions anew, and gradually forms Ideas of such Beings, as are the immediate and direct Object of our Senses. These real Bodies being immediately relative to our Wants, we are nearly concern'd to study them. Mathematical Abstractions facilitate our Knowledge thereof; and prove only useful in Life, by this Application.

And now, after having, in a Manner, exhausted the Properties of figur'd Extension by our geometrical Speculations; we come back, and restore Impenetrability thereto; which constitutes physical Body, and is the last sensible Property our Abstractions divested it of. This new Consideration brings with it that of the Action of Bodies upon one another; for Bodies only act by means of Impenetrability: whence proceed the *Laws of Motion*, and *Equipollency*, which are the Object of *Mechanics*. And hence we carry our Enquiries to the Motion of Bodies, animated even by unknown Forces, or moving Causes; provided the Law whereby these Causes act, is either known, or suppos'd.

The Use of Mathematics is no less considerable in examining the terrestrial Bodies that surround us. All the Properties found in these Bodies have certain Relations to one another, more or less sensible to us. The Knowledge, or Discovery, of these Relations, is almost the only Object we can arrive at, or need propose to ourselves. It is not to be expected that amusing, arbitrary Hypotheses can help us to understand Nature; such Knowledge must be acquired by considering Phaenomena, comparing them together, and reducing, as much as possible, many of them to one, as to a Principle. In fact, the more we diminish the Number of Principles in a Science, the greater Extent we give them. For, as the Object of a Science is necessarily determined, the Principles applied to that Object become so much the more pregnant, the fewer, or more general, they are made. This Art of Reduction, which facilitates the Discovery of Principles, constitutes genuine Theory; and should not be mistaken for the Spirit of building Systems; which is not always the same thing, as will appear hereafter.

Properly speaking, only those Parts that treat of Magnitude, and the general Properties of Extension, as Algebra, Geometry, and Mechanics, can be call'd demonstrative.

> *The Plan of the French Encyclopaedia, or Universal Dictionary of Arts, Sciences, Trades and Manufactures, being An Account of the Origin, Design, Conduct, and Execution of that work.* Translated from the Preface by d'Alembert. London, 1752. Pp. 3, 13-19, 24.

Bibliography:

Traité de dynamique, 1743; Traité de l'équilibre et du mouvement des fluides, 1744; Réflexions sur la cause génerale des vents, 1744; Discours Préliminaire, 1751; Mélanges de littérature, d'histoire et de philosophie, 1752; Eléments de musique théorique et pratique, 1752; Recherches sur différents points important du système du monde, 1754; Essai sur les éléments de la philosophie, 1759.

D. Diderot: *Le rêve d'Alembert*, 1769; J. Bertrand: *D'Alembert*, Paris, 1889; E. Cassirer: *Die Philosophie der Aufklärung*, 1932; M. Muller: *Essai sur la philosophie de J. d'Alembert*, 1926.

Julien Offray De La Mettrie
(1709–1751)

Man as a Machine

The nature of motion is as little known as that of matter, nor have we any means of understanding how motion arises in nature, unless we wish to resuscitate the old and incomprehensible doctrine of 'substantial forms' as the author of *The History of the Soul* has done. Thus I am no more disturbed by the fact that I do not know how inert and simple matter becomes the active and complex matter of the organs, than by the fact that I cannot look at the sun without using a red glass. This is also my emotion when confronting the other wonders of nature, *i.e.*, the production of feelings and thoughts in a being, who once appeared to our limited horizon as just so much dirt.

If I am granted that organized matter is endowed with the principle of motion, which alone distinguishes it from the unorganized (and who could doubt this in the face of so much incontrovertible observation), and that in animals everything depends on this organization as I have already proved at length, this will suffice for solving the puzzle of substances and of man. It is clear that there is but one substance in the world, and that man is its ultimate expression. Compared to monkeys and the cleverest of animals he is just as Huygens' planet clock is to a watch of King Julian. If more wheels and more springs are needed to show the motion of the planets than are required for showing and repeating the hours, and if Vaucanson needed more artistry in producing a flautist than a duck, his art would have been even harder put to produce a 'talker'; and such a machine, especially in the hands of this new kind of Prometheus, must no longer be thought of as impossible.

Equally, nature had perforce to employ more art and technique in producing and maintaining a machine designed to display all the motions of the heart and the mind for a whole century; for if we cannot tell the hours by the pulse, it is nevertheless a barometer of warmth and vitality from which we may conclude as to the nature of the mind. I am certain that I am not mistaken, the human body is a watch, but so astonishing and made with so much art and skill, that if the second-wheel should stop the minute-wheel will continue all the while, just as the quarter-hour wheel and all the others continue in their movements, when the former have become rusty or impaired in any way and interrupted in their course. For it is a fact that obstructions in a few vessels do not suffice for destroying or interrupting the centre of all movements, lying in the heart as the motor part of the machine; on the contrary the fluids whose volumes have become diminished now have a shorter path and traverse it all the more quickly. Furthermore, according as the resistance at the terminals of the vessels increases the power of the heart, these fluids are carried away as if a fresh current were acting upon them. If mere pressure on the visual nerve causes it no longer to transmit the images of objects, why therefore would a loss of vision prevent the use of our auditory faculties, or the loss of the latter by suspending the functions of the *portio mollis* involve the loss of the former? Does it not happen that one man understands but cannot repeat what he has understood (as, for instance, immediately after an apoplectic fit), while another who understands nothing, but whose lingual nerves are not blocked in the brain, can relate like a machine all the dreams that pass through his head? Enlightened physicians cannot be surprised by such phenomena. They know what is essential in human nature, and, by the way, I believe that of two physicians the better and more reliable is always he who is more conversant with the physics or mechanics of the human body and who leaves aside the soul and all the conflicts that this chimera of fools and ignoramuses entails, and concerns himself with pure science alone.

> De la Mettrie: *L'Homme machine*,
> pp. 57 ff. (1748).

Wilhelm Ostwald
(1853–1932)

Lectures on Natural Philosophy

The name Natural Philosophy, with which I have endeavoured to designate the content of our present talks, has a very bad reputation. It recalls a spiritual tendency which reigned in Germany one hundred years ago: its leader was the philosopher Schelling, who, through the force of his personality, had already gained a tremendous influence in his earlier years, and who swayed the thought of his contemporaries. However, this influence was manifest only in the case of Schelling's compatriots—the Germans; Scandinavia, England and France were completely negative in their attitude to 'natural philosophy'.

Even in Germany its absolute reign was but of short duration—twenty years at most. Scientific investigators, for whom this philosophy had been designed particularly, were the first to turn away from it completely, and its subsequent condemnation was to be as passionate as had been its previous exaltation. In order to give some idea of the feelings of some of its former adherents we need only recall those words of Liebig, with which he characterizes his excursion into the realm of natural philosophy: 'I, too, have lived through this period, so rich in words and ideas and so poor in true knowledge and proper studies; it has cost me two valuable years of my life; I cannot describe the shock and the horror, when I awoke to consciousness from this trance.'

Since Natural Philosophy had such effects on its own former disciples, it is not surprising that this mode of thought was soon to disappear from the horizons of scientific investigators. It was replaced by a mechanistic materialist philosophy, which had been developed simultaneously in France and in England. Because of the erroneous notion of its adherents that this philosophy was a conception of reality not involving hypotheses, this intellectual change went hand in hand with a marked aversion for all other viewpoints. The latter were branded as 'speculative', and even today this designation is still considered to be some kind of evil word in scientific circles. It is instructive to note that this aversion is in fact not directed against speculative thought in general, but specifically against such thought as did not belong to the sphere of mechanistic philosophy; mechanistic philosophy was exempt from reproach, for it seemed indistinguishable from immediate scientific results. Thus this anti-philosophical attitude was, at least subjectively, felt to be completely honest.

That 'natural philosophy' was to succumb so quickly to materialism was due simply to practical results. While German philosophers were busy with reflecting and writing on natural phenomena, the representatives of other schools did experiments and were soon able to demonstrate a host of practical results, which were the main reasons for the speedy development of the sciences in the nineteenth century. Philosophers could do nothing in the face of these concrete proofs of superiority. Although they, too, made discoveries, the simultaneous dead-weight of words and unproductive ideas was, according to Liebig, so great that compared with it the actual scientific advances were minimal.

Thus the age of Natural Philosophy has become known as the age of the greatest poverty in German science, and it appears as a presumptuous undertaking for a scientist of the twentieth century to wish to sail under its flag.

However, the name 'natural philosopher' might be given quite another meaning. As in the case of 'natural physician', 'natural singer', *etc.*, 'natural philosopher' could simply refer to those who practise what they have not studied. Nor could I defend myself against this interpretation, for I am a research chemist and physicist and may not call philosophy a subject that I have studied in a general way. Even the 'wild' study of philosophy, which I have made by means of intensive reading in the writings of the philosophers, was so unsystematic that I could by no means claim it was an adequate substitute for disciplined studies. The only extenuating circumstance is the fact that in the course of his work a scientist is unavoidably thrown into contact with the self-same problems that philosophers are concerned with. The mental operations governing scientific labours and bringing them to successful conclusions, are, in their essence, indistinguishable from the investigations and teaching of philosophy. Awareness of this relationship may well have been dimmed in the second half of the nineteenth century, but precisely in our times it had become vitally effective, and everywhere in the camp of scientists there are keen minds who wish to contribute their share to the total philosophic understanding. Thus our age is ready to experience a new development of natural philosophy in both senses of the word, and the great number of those who have gathered today under its banner is evidence that there is something attractive in the fusion of the two concepts of nature and philosophy, and that all of us see a problem here, the solution of which is close to our hearts.

However, the philosophy of a scientific investigator may not claim to be a closed and fully rounded philosophic system.

The creation of such systems is best left to professional philosophers. We are conscious of the fact that our work will at best produce a building, whose construction and internal arrangements will everywhere show signs of the attitudes and habits of thought that we have acquired from our daily concern with particular groups of natural phenomena. While having a constant regard for this personal and occupational style, I must ask you to reflect on what I have to offer you, and all of you are invited to subtract or to add what you consider necessary or desirable.

From the Introduction.

Time, Space and Substance

Thus there are two different groups of reasons, one supporting our adherence to the reality of things as they appear to us, and the other rejecting this attitude. In order to resolve the contradiction contained in this we shall have to prove that both points of view contain lacunæ whose bridging will lead to a fusion. These lacunæ will obviously be found in their respective delimitations of the concept of substance.

The 'substance' of the physics and chemistry of the nineteenth century was given the special name of *matter*. This, as it were, has remained as a residue of the many substances of the eighteenth century. Caloric, the electric and magnetic fluids, light and others, had in the course of time lost their characteristics as substances and had become forced to lead a more spiritual existence as 'forces'. It is difficult to determine unequivocally what is understood by matter today; for if we try to get behind certain definitions, we shall discover that an understanding of this concept is generally taken for granted and that matter is subsequently treated as something self-evident.

However, we should be able to arrive at an approximate delimitation of this concept on the basis of those data on the properties of matter that can be found in any textbook of physics. Here also we shall find signs of developments: while the older textbooks are quite definite on this question, the more recent ones have a tendency to by-pass it as being doubtful and uncertain, and not to discuss it at all. However, by way of synopsis, we can say the following:

All matter has a given quantity; the quantity of matter is usually called *mass*. Furthermore, there are certain qualitative differences in matter which can be explained by the existence of seventy to eighty elements which cannot be changed into one another. Further, matter must be considered as having *extension in space* and *limitation of form*; but the latter is only dependent on matter in certain cases (in solids); in the remainder it is determined by the environment. Again, matter is said to be *impenetrable*, *i.e.*, no two different pieces of matter can exist simultaneously in the same space. Finally matter is said to be *indestructible*.

Sometimes these *essential* properties of matter are distinguished from *general* properties, which, although they are present in all matter, do not essentially belong to its definition. These are the properties of *inertia*, *i.e.*, the ability to retain a given state of motion, *gravity*, *divisibility* and *porosity*. However, there is little agreement which properties are essential and which are general, and frequently no such distinction is made.

This state of affairs in science is nothing less than satisfactory. If you recall your first physics lessons, in which you were told of the basic concepts of physics, you will also recall the hollow feelings that followed all your attempt to get some definite pictures during discussions that must have sent your thoughts spinning in circles. All of us, including the teacher, breathed a sigh of relief when we were allowed to leave these notions and discuss instead the lever, the gravity machine or anything else that was real.

Now such definitions were plainly meant to help us discover and select a series of general properties in the things of the external world. The old concept of matter tried to embrace all things physical, but because of the increasing demand for a definition involving definite limitation, tangibility, and particularly impenetrability, the definition of matter came increasingly to be restricted to objects endowed with mass (in the mechanical sense) and weight. In this way, however, many important phenomena, such as for instance those of light and electricity, are excluded. Apparently these act through the immaterial space between stars, the sun and the earth, without meanwhile adhering to anything material.

Admittedly, attempts have been made to fill the existing hiatus by the assumption of an immaterial substance, *i.e.*, one lacking the above properties but capable of transmitting certain other properties or conditions, the so-called *ether*, and in all textbooks and reviews the physics of matter is treated separately from that of the ether. This is obviously nothing but a stopgap, for all attempts at a valid formulation of the properties of the ether after the analogy of the known properties of matter have led to irresolvable contradictions. Clearly, the ether has crept into science, not because it allows a satisfactory description of facts, but rather because no one has attempted, or known how, to replace it by something better.

If we now continue on the path that has led us this far and set ourselves the task of searching for an honest and coherent description of conditions in the external world, then, particularly in our formulation of the concept of substance, we shall have to rely on experience in the most exact and unprejudiced way possible; for this task implies no less than determining what sort of things have the property of conservation or of durability, and if this should lead us to more than one such concept, to single out that which makes up the constant essence of all external things.

Since the discovery of the *law of the conservation of total mass* at the end of the eighteenth century it has become common practice when speaking of any chemical and physical processes to refer only to those things which can be *weighed* as 'substance' or 'matter'. However, it is by no means only ponderable matter which is conserved under all known conditions. Thus, for instance, in mechanics there is a certain magnitude known as *momentum*, depending both on mass and on velocity, and this also has the property of conservation. Just as in the case of the weight of ponderable matter, we know of no process by which the momentum of a particular system can be changed.

Admittedly it can be altered by allowing collision with other masses having velocities. However, since masses also can neither be created nor destroyed, this apparent exception is due only to the fact that the original momentum of the system was derived without considering the subsequent action of other masses. If we take this factor into consideration from the very start, the law of the conservation of momentum is strictly valid, and we know of no exceptions. The same property of conservation is found in the case of some other imponderable magnitudes known to physics. An example is the *quantity of electricity* which, so long as we bear in mind the sign when adding positive and negative quantities, cannot be changed by any known processes. The result is always an equal quantity of positive and negative electricity whose sum is zero, thus leaving the total quantity unaltered.

Finally, there is yet another magnitude known as *work* or *energy*, whose conservation (in a definite sense) has been known and recognized since the middle of the nineteenth century. Thus it also belongs to those things which can be neither destroyed nor created.

If we test these and other magnitudes governed by the law of conservation we find the following: with the exception of energy, all other concepts whose magnitude obeys the law of conservation can be applied only to limited fields of natural phenomena. *Only energy can be found in all known natural phenomena without exception* or, in other words, *all natural phenomena can be classified under the concept of energy*. Thus the concept is particularly suited to a complete solution of the problems raised by the concept of substance, and not fully resolved by the concept of matter.

Energy is not only present in all natural phenomena but *it is characteristic of them all*. Every process without exception can be described precisely and exhaustively when we state what changes in energy have taken place in time and space. Conversely, the question what are the requisite conditions for any process to occur in the first place, or under what circumstances anything happens at all, can be given a general answer depending on the behaviour of the given energies. Thus even the second aspect required for formulating the most general concept of external things is covered by the consideration of energy. In fact, we can say that *all our knowledge of the external world can be represented in the form of statements about the energies involved*, and thus energy appears from all angles to be the most general concept so far formulated by science.

(Pp. 148 ff.)

Consciousness

There are two possible *effects* of the processes arising out of the action of external energy on our impressions and feelings, and which we may consider to be the creation of nervous energy at the cost of external energy. Either our feelings produce an immediate reaction in such a way that an *action*

MECHANISTIC-MATERIALIST WORLD-VIEW 145

in the most general sense is produced, *i.e.*, the energy of the organism is directed towards the outside, or else there are intermediate transformations of primary nervous energy into other forms. Since even the *release* of an action rests generally upon an intermediate transformation, the latter is the more general phenomenon and we shall discuss it first.

The transformation of nervous energy produced in the sensory apparatus probably takes place in those organs called *ganglion cells*, which are always found at the terminals of every nerve fibre. The process must not be thought of as a mere transformation of energy but as having the character of a *relative release*. In other words, the action of the nervous energy is used to transform by means of a release mechanism the present store of energy, probably of a chemical nature, into fresh nervous energy whose quantity can have various ratios to the quantity of the acting energy according to the property of the transformer. In particular, *habit*, which we have mentioned on many occasions, intervenes in such a way that the releasing energy required for the production of a given quantity of released energy is the less, the more frequently an equal or similar process has previously taken place in the structure concerned.

Now, this newly produced nervous energy is either transferred to the central organ or else it is shifted to those motor organs in which the body develops energies directed towards the outside. In the first case we have consciousness and in the second, unconscious actions or *reflexes*. This interpretation of nervous processes has been so often confirmed by anatomical and physiological discoveries that we may consider it as correct.

I therefore suggest to you that hereafter you consider consciousness as the property of a particular kind of nervous energy, viz., that which takes place in the central organ. The reason that not all nervous energy produces consciousness undoubtedly follows from the fact that when consciousness is excluded during sleep, by anaesthesia or by narcosis, a considerable part of the nervous system, *viz.*, that which regulates the involuntary processes of the body such as the beating of the heart, breathing, digestion, glandular secretion, *etc.*, will continue to work without being disturbed by the absence of consciousness. Also, under such conditions it often happens that we carry out actions that are normally conscious and voluntary.

What, then, is the connection between consciousness and nervous energy? It seems to me that we must look upon this connection as being the closest possible, and I am inclined to assume that consciousness is just as essential a characteristic of the central organ as space is of mechanical energy and time of the energy of motion. This will become clearer if we recall the point of departure of our discussion, where we pointed out that our entire knowledge of the external world is due to processes in our consciousness. From all the factors common to our experiences, we have found that the *concept of energy* is the most general and, according to the nature of these experiences and their common relations, we have distinguished different kinds of interchangeable energy. Thus no one can accuse us of inconsistency when we try to connect the source of all these notions, our consciousness itself, with this most general of concepts, and if we say with Kant that all our notions of the external world are subjective inasmuch as we assimilate only such activities of the external world as correspond with the nature of our consciousness. The fact that all external events can be represented as energetic processes is best explained by the assumption that *our conscious processes are themselves energetic processes which stamp this property of theirs upon all external experiences.* All I ask you here is merely to consider this notion as an attempt to arrive at a unified conception of the world, an attempt which we must always make when it is a question of grasping a new field or

MECHANISTIC-MATERIALIST WORLD-VIEW 147

finding new paths for understanding an old one. Experimentally, such notions are tested by developing all their consequences and then comparing known facts with these consequences.

Now all psychologists agree that energy processes accompany all mental and particularly conscious processes and that all thought, sensation and volition involves an expenditure of energy. Now, the theory of *psycho-physical parallelism* has been considered adequate for an understanding of this fact. In its older form, as given by Spinoza, this theory states that mental and physical processes are but different aspects of the same factual event, and according as we look upon substance from the point of view of extension (*i.e.*, physically) or of thought (*i.e.*, psychologically) we obtain the one or the other kind of phenomenon. The more recent theory of psycho-physical parallelism rejects this conception as unscientific, and replaces it by a process involving two simultaneous and parallel causal sequences, which, because of the incompatibility of their members, can never become one. I find it difficult to discover any difference between this principle and Leibnitz's pre-stabilized harmony, apart from the introduction by Leibnitz of the hypothetical concept of a *monad*, and that of matter by modern exponents. Even the adherents of this theory admit that all is not well with their parallelism and hope that further developments may remove the difficulties. However, in view of the progress made in all spheres of knowledge it is hardly probable that the unifying idea is to be found in the field of metaphysics alone, and not in the field of science to which both physiology and psychology belong. Time has shown that all assertions of the impossibility of progress along lines of general development have always been wrong.

If we wish to find the reason why so difficult an idea as that of independent parallelism arose in the first place, we must look for it in *mechanistic materialism*. Even Leibnitz had already realized this connection, and in our times Dubois-Reymond has illustrated it with his postulation of 'ignorabimus'. Leibnitz points out that if, keeping all its other properties the same, we could imagine a human brain so great that we could look into it and walk about in it 'like a mill', and if we could understand completely all the mechanisms of the atoms in the brain, we should still see nothing but movable atoms, and nothing of the thought corresponding to these motions. Similar ideas were developed by Du Bois-Reymond in his lecture on the limits of our understanding of nature. Knowledge of the masses, velocities, positions and forces of the molecules in the brain would have to be of 'astronomical' proportions, or as he puts it: 'Now as regards the mental processes themselves, it is quite clear that even with an astronomical knowledge of the organ of the mind we should understand them just as little as we do now. Even if we had this knowledge we should not find them less incomprehensible. An astronomical knowledge of the brain, *i.e.*, the most complete knowledge of it that we can obtain, will reveal in it nothing but moving matter. However, no amount of ordering or moving of material particles can ever allow us to build a bridge into the realm of consciousness.'

I know of no more convincing proof for the validity of a philosophy of energetics than the fact, here demonstrated, that it robs this old problem of all its terror, since the difficulty only arose from the fact that Leibnitz and Dubois-Reymond assumed with Descartes that the physical world consisted of nothing but movable matter. Obviously in such a world there is no room for thought. We, who consider energy as the ultimate reality, are not troubled by such impossibilities. We have seen that there are no contradictions in deriving nerve conduction from energy processes, and we have also seen that those nervous processes associated with consciousness

MECHANISTIC-MATERIALIST WORLD-VIEW 149

are always connected with unconscious ones. I have gone to great pains to look for any absurdities or impossibilities in the assumption that consciousness is governed by particular kinds of energy. I have been unable to discover any such thing. Very soon, when investigating the most important phenomena of consciousness, we shall become convinced that they are all governed by energy, and I have no more difficulty in imagining that kinetic energy governs *motion* than I have in thinking that the energy of the central nervous system governs *consciousness*.

At the same time we realize that the energy connected with consciousness is the highest and rarest form of energy known to us. It is only produced in specially developed organs, and even in the brains of different people we can find the greatest differences in the quantity and effectiveness of this energy. We must not be surprised that such energy is only produced under special circumstances. Similarly, relatively few of the great number of known crystals are suitable for the production of electrical energy through pressure, *i.e.*, only those which are uniaxial. The radiation of uranium and some other elements which has been investigated in our time is an energy process that occurs even more infrequently, and the conditions of whose production are even more limited.

We can similarly avoid another great difficulty. If our experience has shown us that, in man, mind is always associated with the 'matter' of his brain, then there is no reason why mind should not be associated with other matter, since the elements carbon, hydrogen, oxygen, nitrogen and phosphorus in the brain are exactly the same as those existing elsewhere. Although by means of metabolic exchange they are constantly being replaced by others, their action on the brain is not affected. Thus, if mind is a property or an action of the matter in the brain, then, according to the law of the conservation of matter, this property must unconditionally be attributed to all the atoms postulated by mechanistic philosophy, and a stone, a table or a cigar are as endowed with mind as is a tree, an animal or a man. In fact, this conclusion is inevitable if we grant the assumption on which it is based, and in more recent philosophic literature we find that it is either accepted or at least considered appropriate; otherwise a definite and unbridgeable dualism of mind and matter must be postulated to avoid it.

This difficulty, too, disappears in the case of energy. While matter obeys the law of the conservation of elements so that in a given space the quantity of oxygen, nitrogen, *etc.*, either in compounds or in the free state, cannot be changed by any known process, it is generally possible to transform a given quantity of energy into another without any measurable part of the former remaining. Thus experience is by no means opposed to the notion that particular types of energy also need certain conditions for their production, and that given quantities can be completely transformed into other forms. This is the case with mental energy, *i.e.*, with unconscious and conscious nervous energy.

If a conception of the mind based on energetics thus recommends itself by its ability to resolve great difficulties, a task that had challenged the intelligence of many centuries, there still remains the most important task of testing whether conscious mental activity can also be fitted into the framework of the conception of energy without leading to contradictions. I believe that I can give an affirmative reply. I must stress that this is only a tentative opinion; for a scientific decision of the matter a great deal of work of the most difficult kind is still needed. Yet the following reflections seem to assure energetics of a hopeful future.

The new theory of psycho-physical parallelism starts from the premise that for each mental event there exists a corresponding physical event, and in so far as we have been able to test this assumption at all it has been confirmed. Similarly, materialists assume that the mind is nothing but an effect of matter, and in support of this wide-spread philosophy we are given a great number of experimental facts. Energetics can mobilize both armies on its side since both 'physical event' and 'effect of matter' in our sense are nothing but the transformation of energy. The difference is due only to the untenable assumption that matter is the ultimate concept of reality. If we abandon it, the front changes and all the proofs produced by both camps serve the interests of the conception of energetics.

Bibliography:

Lehrbuch der allgemeinen Chemie, 2 vols., 12th ed., 1910–1911; *Die Überwindung des wissenschaftlichen Materialismus*, 1895; *Vorlesungen über Naturphilosophie*, 1902; *Die Harmonie der Farben*, 5th ed., 1923; *Die Harmonie der Formen*, 1922; *Lebenslinien*, 3 vols., 2nd ed., 1932–33.

Translations:

Conversations on Chemistry (translated by E. C. Ramsay); *Colour Science* (translated by J. S. Taylor); *Outlines of General Chemistry* (translated by J. Walker); *Principles of Inorganic Chemistry* (A. Findlay); *Natural Philosophy* (T. Seltzer).

V. Delbos: *Wilhelm Ostwald et sa philosophie*, 1916; A. Mittasch: *Wilhelm Ostwald's Auslösungslehre*, 1951; G. Ostwald: *Wilhelm Ostwald, mein Vater*, 1953; J. S. Taylor: *A Simple Explanation of the Ostwald Colour System*.

6

THE CRISIS OF THE MECHANISTIC-MATERIALIST CONCEPTION

IN THE first parts of this section we have dealt with the beginnings of modern scientific thought and the rise of the mechanistic-materialist philosophy. These were illustrated by lengthy quotations from classical writers who were both the originators and also the promoters of this development. For reasons of space we shall limit this part to a lengthy extract from the writings of Louis de Broglie, to whom we owe a model account of what has caused the crisis in mechanistic-materialist thought.

However, by way of preface we shall also quote the introduction to the *Principles of Mechanics* (1876) by Heinrich Hertz (1857–1894), for here it emerges clearly how physics began to remember once more that a natural science is one *whose propositions on limited domains of nature can have only a correspondingly limited validity; and that science is not a philosophy developing a world-view of nature as a whole or about the essence of things*. Hertz points out that propositions in physics have neither the task nor the capacity of revealing the inherent essence of natural phenomena. He concludes that physical determinations are only pictures, on whose correspondence with natural objects we can make but the single assertion, *viz.*, whether or not the *logically* derivable consequences of our pictures correspond with the empirically observed consequences of the phenomena for which we have designed our picture. In other words, the hypothetical picture of a causal relationship with which we invest natural phenomena must prove its usefulness in practice. The criteria for assessing the suitability of a picture are that (1) it must be *admissible*, *i.e.*, correspond with our laws of thought; (2) it must be *correct*, *i.e.*, agree with experience; (3) it must be *relevant*, *i.e.*, contain the maximum of essential and the minimum of superfluous or empty relations of the object.

Here already we get a foretaste of the essential insight of modern physics, stated with such impressive brevity by Eddington: 'We have found that where science has progressed the farthest, the mind has but regained from nature that which the mind has put into nature. We have found a strange footprint on the shores of the unknown. We have devised profound theories, one after another, to account for its origin. At last, we have succeeded in reconstructing the creature that made the footprint. And Lo! it is our own.'

Heinrich Hertz
(1857–1894)

Introduction to the Principles of Mechanics

The most direct, and in a sense the most important, problem which our conscious knowledge of nature should enable us to solve is the anticipation of future events, so that we may arrange our present affairs in accordance with such anticipation. As a basis for the solution of this problem we always make use of our knowledge of events which have already occurred, obtained by chance observation or by prearranged experiment. In endeavouring thus to draw inferences as to the future from the past, we always adopt the following process. We form for ourselves images or symbols of external objects; and the form which we give them is such that the necessary consequents of the images in thought are always the images of the necessary consequents in nature of the things pictured. In order that this requirement may be satisfied, there must be a certain conformity between nature and our thought. Experience teaches us that the requirement can be satisfied, and hence that such a conformity does in fact exist. When from our accumulated previous experience we have once succeeded in deducing images of the desired nature, we can then in a short time develop by means of them, as by means of models, the consequences which in the external world only arise in a comparatively long time, or as the result of our own interposition. We are thus enabled to be in advance of the facts, and to decide as to present affairs in accordance with the insight so obtained. The images which we here speak of are our conceptions of things. With the things themselves they are in conformity in *one* important respect, namely, in satisfying the above-mentioned requirement. For our purpose it

MECHANISTIC-MATERIALIST CONCEPTION 155

is not necessary that they should be in conformity with the things in any other respect whatever. As a matter of fact, we do not know, nor have we any means of knowing, whether our conceptions of things are in conformity with them in any other than this *one* fundamental respect.

The images which we may form of things are not determined without ambiguity by the requirement that the consequents of the images must be the images of the consequents. Various images of the same objects are possible, and these images may differ in various respects. We should at once denote as inadmissible all images which implicitly contradict the laws of our thought. Hence we postulate in the first place that all our images shall be logically permissible—or, briefly, that they shall be permissible. We shall denote as incorrect any permissible images, if their essential relations contradict the relations of external things, *i.e.*, if they do not satisfy our first fundamental requirement. Hence we postulate in the second place that our images shall be correct. But two permissible and correct images of the same external objects may yet differ in respect of appropriateness. Of two images of the same object, that is the more appropriate which pictures more of the essential relations of the object—the one which we may call the more distinct. Of two images of equal distinctness the more appropriate is the one which contains, in addition to the essential characteristics, the smaller number of superfluous or empty relations—the simpler of the two. Empty relations cannot be altogether avoided: they enter into the images because they are simply images—images produced by our mind and necessarily affected by the characteristics of its mode of portrayal.

The postulates already mentioned are those which we assign to the images themselves: to a scientific representation of the images we assign different postulates. We require of this that it should lead us to a clear conception of what properties are to be ascribed to the images for the sake of permissibility, what for correctness, and what for appropriateness. Only thus can we attain the possibility of modifying and improving our images. What is ascribed to the images for the sake of appropriateness is contained in the notations, definitions, abbreviations, and, in short, all that we can arbitrarily add or take away. What enters into the images for the sake of correctness is contained in the results of experience, from which the images are built up. What enters into the images, in order that they may be permissible, is given by the nature of our mind. To the question whether an image is permissible or not, we can without ambiguity answer yes or no; and our decision will hold good for all time. And equally without ambiguity we can decide whether an image is correct or not; but only according to the state of our present experience, and permitting an appeal to later and riper experience. But we cannot decide without ambiguity whether an image is appropriate or not; as to this, differences of opinion may arise. One image may be more suitable for one purpose, another for another; only by gradually testing many images can we finally succeed in obtaining the most appropriate.

Those are, in my opinion, the standpoints from which we must estimate the value of physical theories and the value of the representations of physical theories. They are the standpoints from which we shall here consider the representations which have been given of the Principles of Mechanics. We must first explain clearly what we denote by this name.

Strictly speaking, what was originally termed in mechanics a principle was such a statement as could not be traced back to other propositions in mechanics, but was regarded as a direct result obtained from other sources of knowledge. In the course of historical development it inevitably came to pass that propositions, which at one time and under special circumstances were rightly denoted as principles, wrongly retained these names. Since Lagrange's time it has frequently been remarked that the principles of the centre of gravity and of areas are in reality only propositions of a general nature. But we can with equal justice say that other so-called principles cannot bear this name, but must descend to the rank of propositions or corollaries, when the representation of mechanics becomes based upon one or more of the others. Thus the idea of a mechanical principle has not been kept sharply defined. We shall therefore retain for such propositions, when mentioning them separately, their customary names. But these separate concrete propositions are not what we shall have in mind when we speak simply and generally of the principles of mechanics: by this will be meant any selection from amongst such and similar propositions, which satisfies the requirement that the whole of mechanics can be developed from it by purely deductive reasoning without any further appeal to experience. In this sense the fundamental ideas of mechanics, together with the principles connecting them, represent the simplest image which physics can produce of things in the sensible world and the processes which occur in it. By varying the choice of the propositions which we take as fundamental, we can give various representations of the principles of mechanics. Hence we can thus obtain various images of things; and these images we can test and compare with each other in respect of permissibility, correctness, and appropriateness.

From the translation by D. E. Jones.

Bibliography:

Gesammelte Werke, 3 vols., 1894–1895 (*Electric Waves and the Principles of Mechanics*, translated by D. E. Jones).

M. Planck: *Heinrich Hertz*, 1894; Johanna Hertz: *Heinrich Hertz, Erinnerungen, Briefe, Tagebücher*, 1927; J. Zenneck: *Heinrich Hertz*, 1929.

Louis de Broglie
(1894)

The Progress of Contemporary Physics

Like all the other natural sciences, Physics advances by two distinct roads. On the one hand it operates empirically, and thus is enabled to discover and analyse a growing number of phenomena—in this instance, of physical facts; on the other hand it also operates by theory, which allows it to collect and assemble the known facts in one consistent system, and to predict new ones for the guidance of experimental research. In this way the joint efforts of experiment and theory, at any given time, provide the body of knowledge which is the sum total of the Physics of the day.

At the beginning of the development of modern Science, it was naturally enough the study of the physical phenomena which we observe immediately around us that first drew the attention of physicists. Thus the investigation of the equilibrium and the motion of bodies led to the development of the branch of Physics—today an independent study—known as mechanics. Similarly research into the phenomena of sound led to acoustics, while optics was created by collecting the phenomena of light and forming them into one system.

The great task and the splendid achievement of nineteenth-century Physics consisted in thus increasing the exactness and range—in every direction—of our knowledge of the phenomena taking place on the human scale. Not only did it continue to develop mechanics, acoustics and optics—the leading branches of classical Science—but it also created on every side new sciences possessing innumerable aspects, such as thermodynamics and the science of electricity.

The mastery of the vast sphere of facts covered by these

various branches of Physics has enabled both abstract students and technical workers to draw thence a great number of practical applications. The inventions—ranging from the steam engine to wireless telephony—derived from the nineteenth-century advance of Physics, the benefits of which we enjoy today, are innumerable; and these inventions play so important a part, directly or indirectly, in the everyday life of each of us that it would be wholly superfluous to enumerate them.

In this way, then, nineteenth-century Physics succeeded in achieving the complete domination of the phenomena we observe around us. No doubt research into these phenomena can still lead to the knowledge of many further facts and to new applications; yet it appears that in this sphere the essential work has now been completed. And, in fact, during the last thirty or forty years the attention of pioneers in Physics has been turning increasingly towards more subtle phenomena, which could be neither discovered nor analysed without an extremely refined experimental technique: molecular, atomic and infra-atomic phenomena. The fact is that in order to satisfy human curiosity it is not enough to know the behaviour of material bodies taken as wholes, or in their manifestations *en masse*, or to grasp the reactions between Light and Matter when observed on the macroscopic scale: what is required is to descend to individual details, to attempt the analysis of the structure of both Matter and Light, and to specify the elementary processes which in their totality constitute the macroscopic phenomena. It is a difficult inquiry, and for its success an extremely delicate experimental technique is required, capable of discovering and recording exceedingly subtle events, and of measuring exactly magnitudes vastly smaller than those occurring in our everyday experience. Still further, bold theories are required, based on the highest branches of Mathematics and prepared to make use of entirely novel similes and concepts. Hence we can infer the amount of ingenuity, patience and talent needed for the formulation and advancement of this atomic Physics.

On the experimental side, then, the progress made has been characterized by a daily growing knowledge of the ultimate constituent entities of Matter and of the phenomena connected with the existence of these ultimate constituent entities.

Chemistry had long assumed that material substances are composed of atoms; and the actual investigation of the properties of material substances shows them to be divided into two classes: compound substances, which can be reduced to simpler ones by appropriate methods; and the simple substances themselves—the chemical elements—which resist any attempt at such reduction. In the next place, the study of the quantitative laws, in accordance with which the simple substances combine to form compounds, led chemists during last century to adopt the following hypothesis:

'A simple substance is supposed to be formed of small particles, all identical with each other, called the atoms of this element; compounds, on the other hand, are supposed to be formed of molecules resulting from the combination of a number of the atoms constituting the simple substances.' According to this hypothesis, therefore, a composite substance is broken up by reducing it to the elements of which it is composed, which means that its molecules are disintegrated and the atoms which they contain set free. The number of these simple substances known today is eighty-nine, but it is believed that their total number is ninety-two (or possibly ninety-three). All material substances, therefore, are regarded as constructed from ninety-two different kinds of atoms.

The Atomic Theory not only succeeded in introducing order into Chemistry; it also extended into the domain of

Physics. For if material substances are composed of molecules and atoms, then their physical properties must be capable of explanation in terms of their atomic structure. The properties of the various gases, for example, must be explicable on the assumption that a given gas consists of an immense number of molecules or atoms in rapid motion; the pressure of a gas on the wall of the containing vessel will then be due to the impacts of the molecules against the wall, while the temperature of the gas will be the measure of the average of the motions of the molecules, which increase as the temperature rises. During the second half of the nineteenth century, this view of the structure of gases was developed under the name of the Kinetic Theory of Gases, and it enables us to understand the origin of the laws governing the behaviour of gases as discovered experimentally. For if the Atomic Theory is correct, then the properties of solids and liquids must be capable of interpretation on the assumption that, in the solid and the liquid states, the molecules or atoms are much closer to each other than in the gaseous state. Thus there is an interplay of considerable forces between atoms and molecules in these states, and these should account for such characteristic properties of solids and liquids as incompressibility and cohesion. The Atomic Theory of Matter, again, has been confirmed by brilliant direct experiments such as those of Jean Perrin, by means of which it has been possible to measure the weights of different kinds of atoms and to find their number per cubic centimetre.

Without entering further into the evolution of the Atomic Theory I shall confine myself to recalling that in Physics, just as in Chemistry, the theory which assumed that all substances consist of molecules, which in turn consist of different combinations of elementary atoms, proved very fruitful in practice and can hence be fairly regarded as a useful statement of the actual facts. But physicists did not rest content at this point.

They wished further to discover the structure of the atoms themselves, and to understand the *differentiae* subsisting between the atoms of the different elements; and in this research they were aided by our increasing knowledge of electrical phenomena. When these phenomena first began to be investigated it appeared expedient to treat, for example, the electric current passing through a metallic wire as though it were tantamount to the passing of an 'electric fluid' through the wire. But we know that there are two kinds of electricity —positive and negative. Hence it is natural to assume that there are two fluids: the positive and the negative electrical fluid. These fluids, again, can be imagined in two different ways: we may imagine that they consist of a substance uniformly occupying the whole of the space where the fluid is; or we may imagine that they consist of clouds of little corpuscles each of which is a minute sphere of electricity. Experiment, however, has decided in favour of the second view, and some thirty years ago it showed that negative electricity consists of minute corpuscles which are all identical, and have a mass and an electric charge of extremely small dimensions called electrons. These have been successfully segregated from Matter in bulk, and their behaviour when moving in empty space has been observed; and it has been found that in fact they move in the way in which small particles, electrically charged, ought to move in accordance with the Laws of classical Mechanics; while by observing their behaviour in the presence of electrical or magnetic fields it has proved possible to measure both their charge and their mass—which, I repeat, are extremely small. The demonstration of the corpuscular structure of positive electricity, on the other hand, is less direct; nevertheless physicists have come to the conclusion that positive electricity, too, is subdivided into corpuscles which are identical with each other, today known as protons.

MECHANISTIC-MATERIALIST CONCEPTION 163

The proton has a mass which, though still extremely small, is nearly 2,000 times greater than that of the electron, a fact indicative of a curious asymmetry between positive and negative electricity. The charge of the proton, on the other hand, is equal to that of the electron in absolute value, but of course bears an opposite sign, being positive and not negative.

Electrons and protons, then, have extremely small mass. This mass, however, is not equal to zero, and a really vast number of protons and electrons may make up a fairly considerable total mass. Hence it is tempting to assume that all material substances—whose essential characteristics consist in the fact that they possess weight and inertia, in other words, that they have mass—consist in the last analysis exclusively of vast numbers of protons and electrons. On this view the atoms of the elements, which are the ultimate fabric of which material substances are composed, should themselves consist of electrons and protons; and the ninety-two kinds of different atoms, already referred to, of which the ninety-two elements are composed, should be ninety-two different combinations of electrons and protons. The idea that atoms consist of protons and electrons was next formulated in more exact terms as the result of the experiments of the great British physicist, Lord Rutherford, and of the theoretic work of the Danish scientist, Niels Bohr. The atom of a simple substance was thus shown to consist of a central nucleus, having a positive charge equal to a whole number N times as great as the charge of the proton, and of N electrons gravitating around the nucleus. The entire system, therefore, is electrically neutral, and the nucleus itself is doubtless formed of protons and electrons in the way which we shall see in greater detail below. Almost the entire mass of the atom is concentrated in the nucleus, for the latter contains protons, and these in turn are very much heavier than electrons. The Hydrogen atom is the simplest of the atoms, and consists of a nucleus formed by a single proton around which a single electron revolves. The atom of one element is differentiated from that of another by the number N of positive elementary charges which the nucleus carries. Simple substances can thus be arranged in a series according to the ascending value of the number N, beginning with Hydrogen (N = 1) and ending with Uranium (N = 92). It has been found that this way of classifying substances agrees with that which had been inferred from the value of their atomic weights and from their chemical properties, an arrangement known as Mendeleyeff's classification, after the name of the Russian chemist who first proposed it.

I cannot here explain in detail why the idea that the atom is a kind of minature solar system, with the nucleus for sun and the electrons for planets, has met with so much favour from physicists. I will only say that it has provided an interpretation, not only of the chemical properties of simple substances, but also of several of their physical properties, such as the light rays which they can emit in certain circumstances, for example when incandescent.

One point, however, must be noted. In order to achieve a satisfactory formulation of the theory that the atom is equivalent to a kind of solar system, Bohr had to import a foreign idea, borrowed from the Quantum Theory previously worked out by Planck. I said above that in the experiments in which we are able to follow the motion of an electron the latter behaves like a small corpuscle of very slight mass, and that its motion can be predicted by applying the Laws of classical Mechanics. Let us consider, however, the motion of an electron along a particularly short trajectory. We cannot follow this motion by actual observation; but Bohr has done so in imagination, in order to calculate the characteristic properties of the atom when treated, as he treats it, as a planetary system. Planck, indeed, was himself the first to

MECHANISTIC-MATERIALIST CONCEPTION 165

find that this motion cannot conform exactly to the laws of classical Mechanics. For among the totality of movements which classical Mechanics regards as possible, those which the electron can in fact execute form only a fraction: and this latter privileged group have been called 'quantized'. Bohr therefore, in his theory of the parallelism between the atom and the solar system, has been forced to incorporate Planck's idea and has found that, in fact, the planet-electrons can only have quantized motion; and it is this fact which, in a measure, has provided the key to all the properties of the atoms.

Let us now sum up. Investigation of the properties of material substances has led physicists to treat Matter as consisting solely of small corpuscles, called electrons and protons. Various combinations of these corpuscles constitute the atoms of the ninety-two simple substances which form the raw material of the molecules from which compounds are built up. Such was the conclusion reached some twenty years ago; but we shall shortly see that conditions have since become far less simple; for the moment, however, we must leave the subject of Matter and turn to that of Light.

When Light reaches us from the sun or the stars it comes to the eye after a journey across vast spaces void of Matter. It follows from this that Light can cross empty space without difficulty, wherein it differs for example from sound, since it is not bound up with any motion of Matter. Hence a description of the physical world would remain incomplete unless we were to add to Matter another reality independent of it. This entity is Light.

Now what is Light? What is its structure?

The ancient philosophers, and many scientists until the beginning of last century, maintained that Light consisted of minute corpuscles in a state of rapid motion; and the fact that Light travels in straight lines under ordinary conditions, and its reflection in a mirror, are explained at once by this hypothesis.

But the corpuscular Theory of Light was abandoned entirely, about a century ago, in consequence of the work of the English physicist Young and, even more, of the research of a brilliant French scientist, Augustin Fresnel. Actually, Young and Fresnel discovered a whole set of luminous phenomena—those of interference and of diffraction—which could not be accounted for at all on the corpuscular Theory, while the adoption of another concept—the Wave Theory of Light—accounts both for the classical phenomena of motion in a straight line, of reflection and refraction, and also for the phenomena of interference and diffraction. Fresnel's demonstration of all this was an admirable one.

The Wave Theory of Light—which had previously been adopted by the Dutch scientist Christian Huygens and other far-sighted thinkers—holds that the propagation of Light should be compared to that of a wave in an elastic medium, like the ripples which travel on the surface of a sheet of water when a stone is thrown in. And since Light moves in empty space, Fresnel assumed the existence of a particularly subtle medium—the Ether—supposed to penetrate all material substances, to fill empty space and to act as vehicle for the light-waves.

Let me now explain the way in which a wave is to be imagined. When a wave moves freely it may be compared to a succession of ripples in water, their crests being separated by a constant distance known as the wave-length. The entire group of these ripples moves in the direction of propagation with a certain velocity: that at which the wave advances. For light-waves in empty space this velocity has been shown by experiments, made after Fresnel's death, to be 300,000 kilometres per second. The different waves with their crests and troughs pass a given point in space in succession; and at this point, whatever magnitude it is that is travelling in the form of waves must pass through a periodic variation, the period itself being obviously equal to the time elapsing between the passing of two consecutive crests.

The three magnitudes—the velocity, length and frequency of the wave—are not independent of each other, the frequency being obviously equal to the velocity divided by the wave-length.

We have seen how a wave advances in a region where there is nothing to interfere with its propagation. But conditions are different when the wave meets with an obstacle in the course of its journey; for example if it meets with a surface which stops or reflects it, or again if it has to pass through an aperture in a screen, or if it meets particles of matter which diffract it. In such a case the wave will be deformed and turned back on itself, with the result that instead of a simple wave we shall have a multiplicity of simple, but superimposed waves; and then the resulting type of vibration, at any given point, depends on the way in which the simple superimposed waves tend to reinforce or to enfeeble each other. If there is an additive effect as between the various simple waves, or if they are in phase, as it is called, then the resulting vibration will be one of great intensity; while if their phases are in opposition, the resulting vibration will be weak or even non-existent. To sum up, the existence of obstacles interfering with the propagation of a wave brings about a complicated distribution of the various intensities of vibration, the distribution depending in the main on the wave-length of the wave which meets with the obstacle in question. Of this type are the phenomena of interference and diffraction.

If now we adopt the idea that Light consists of waves, we are led to expect that if there is an obstacle in the free path of a beam of light, then phenomena of interference and diffraction will occur; and Young, followed by Fresnel, showed that under these conditions Light does in fact present phenomena of interference or diffraction; while Fresnel proved, still further, that the Wave Theory of Light affords an adequate explanation of all the observed phenomena in all their details. From that moment, and throughout the rest of last century, the pure Wave Theory of Light was accepted without demur.

There exist, of course, various kinds of light, each corresponding to some definite 'colour'. The white light radiated, for example, by an incandescent body like the filament of an electric lamp, is formed by the superposition of a continuous sequence of simple forms of light whose colours pass by imperceptible gradations from violet to red, thus forming the spectrum. Hence the Wave Theory of Light is naturally led to associate with each kind—with each component of the spectrum—one given wave-length; in other words, one given wave-length corresponds with each colour. Since the interference phenomena depend on the wave-length, they enable us to measure the wave-lengths corresponding to the various colours of the spectrum; and it has proved possible in this way to ascertain that the wave-length varies progressively and continuously from the violet end of the spectrum, where it has the value of 4/10,000ths of a millimetre, to the red end, where it reaches 8/10,000ths.

We have seen, then, that some thirty years ago, no doubts were entertained but that Light, and other kinds of rays, were pure wave phenomena. Since then, however, phenomena due to radiation hitherto unknown have been discovered; and these phenomena apparently can be explained only by a corpuscular theory. The most important of these is the photo-electric effect: when (that is to say) a piece of matter, of metal for example, is illuminated, it is often observed to expel electrons in rapid motion; and observation of this phenomenon has shown that the velocity of the expelled electrons depends solely on the wave-length of the rays falling on the substance, and on the properties of this. But it depends in no way on the intensity of these incident rays: what does solely depend on this intensity is the number of electrons expelled. Further, the energy of the electrons expelled varies inversely with the length of the wave which falls on the substance in question. Consideration of this phenomenon led Einstein to grasp the fact that its explanation demanded a return, at least to some extent, to the theory of the corpuscular structure of radiation. He assumed therefore that rays are composed of corpuscles, the energy of which varies inversely with the wave-length, and has shown that the laws of the photo-electric effect follow easily once this hypothesis is adopted.

At this stage, however, physicists were in a state of no small difficulty. For, on the one hand, they had the group of diffraction and interference phenomena, which show that Light consists of waves; while on the other hand, there were the photo-electric effect and other more recently discovered phenomena, showing that Light consists of corpuscles—of photons, as they are now called.

The only way of escaping from this difficulty, then, is to assume that the wave aspect of Light, and its corpuscular aspect, are as it were two different aspects of the same underlying reality. Thus whenever a ray exchanges energy with Matter, the exchange can be described on the assumption that a photon is absorbed (or emitted) by Matter; on the other hand, if we wish to describe the motion *en masse* of light-corpuscles in space, then we must fall back on the assumption that propagation of waves is taking place. An elaboration of this idea leads to the further assumption that the density of the cloud of corpuscles, which is associated with a light-wave, is at any given point proportional to the intensity of this wave. In this way, therefore, a sort of synthesis of the two ancient rival theories is reached, so that we are enabled to explain interference phenomena as well as the photo-electric effect; but the capital interest of this synthesis consists of the fact that it shows us that, in the world of Nature, waves and corpuscles are closely interconnected—at any rate in the case of Light. And if this interconnection exists, may one not assume that it exists also for Matter? For the entire work of physicists had thus far tended to reduce Matter to a stage where it was no more than a vast collection of corpuscles. But if a photon cannot be separated from the wave which is bound up with it, then surely in the same way we are bound to assume that corpuscles of Matter are in their turn, too, universally associated with a wave. And this, in fact, is the chief question with which today we have to deal.

Let us assume, then, that corpuscles of Matter—electrons, for example—are universally accompanied by a wave. Between the corpuscle and the wave there is an intimate lien; hence the motion of the corpuscle, and that of the wave, are not independent of each other, so that a connection can now be established between the mechanical properties of the corpuscle—its momentum and its energy—on the one hand, and the characteristic values of the wave with which it is associated—its length and the velocity with which it travels—on the other. Thus on the assumption of the interconnection between the photon and its associated wave this parallelism can in fact be established: and this theory of the interconnection between the corpuscles of Matter and their associated waves is known today under the name of Wave Mechanics.

When the wave associated in this way with a corpuscle is moving freely in a region whose dimensions are great as compared with the length of the wave, the New Mechanics assigns to the corpuscle associated with the wave the motion determined by the laws of classical Mechanics. This applies particularly to the motion of electrons which we can observe directly; and this explains why observations of the large-scale motions of electrons had led to their being regarded as simple corpuscles. But there are certain cases where the laws of classical Mechanics fail to describe the motion of corpuscles. The first case is one where the propagation of the associated wave is confined to a region in space having dimensions of the same order as the wave-length; and this is the case of the electrons within the atom. Here the wave associated with an electron is forced to take the form of a stationary wave, similar to the stationary elastic waves found in a cord fixed at each end, or to the stationary electric waves which may be set up in the antenna of a wireless installation. Now theory shows that these stationary waves must have certain quite definite lengths, and that in the associated electron certain equally definite energies correspond to these wave-lengths; still further, these definite states of energy in turn correspond to the states of 'quantized' motion introduced into his theory by Bohr. This also furnishes an explanation for a fact which had hitherto remained extremely mysterious—the fact, namely, that quantized motion is the only type of which the electron contained within the atom is capable.

There is still another case where the electron cannot move in accord with the classical laws of Mechanics—namely where the associated wave meets with obstacles in the course of its advance. In such a case interference takes place, and the motion of the corpuscles, in relation to the motion which classical Mechanics would predict, is somewhat modified; so that to form an idea of what must then occur we may follow the analogy with rays. Let us assume, therefore, that we direct a ray of known wave-length on to an apparatus designed to give rise to interference. Since we know that the rays consist of photons, we can say that we are launching a swarm of photons upon the apparatus; and in the region where interference occurs, the photons are distributed in such a way that they are concentrated at those points where the intensity of the associated wave is greatest. Let us now suppose, still further, that we direct on the same apparatus, not a ray, but a beam of electrons having an associated wave of the same wave-length as in the previous ray. In such a case the wave will interfere as before, since it is the wave-length which controls interference phenomena. It would then be natural to assume that the electrons will be concentrated at the points of greatest intensity of the wave; in other words, that in this second experiment the electrons will be spatially distributed in the same way as that in which the photons were distributed in the first. If then it can be shown that such is in fact the case, the existence of the wave associated with the electrons will also have been demonstrated, and it will thus be possible to check the precision of the formulae of Wave Mechanics.

Now according to Wave Mechanics, a wave is associated with electrons moving with velocities usually realized experimentally, the length of the associated wave being of the same order as that of X-rays, viz. 1/10,000,000th of a millimetre. In order, therefore, to demonstrate electron-waves, we must try to produce by their means interference phenomena analogous to those obtained with X-rays; and phenomena of this type were in fact obtained—first, in 1927, by Davisson and Germer in the United States, and, later, by a great number of experimenters, among whom may be mentioned G. P. Thomson in England and Ponte in France. I shall not describe their experiments, but confine myself to saying that they ended with the complete verification of the formulae of Wave Mechanics.

These brilliant experiments have thus proved that the electron is not merely a simple corpuscle; in one sense it is at once a corpuscle and a wave. The same conclusion—as has been proved by still more recent experiments—applies to the proton. Thus we see that Matter, as well as Light, consists of both waves and corpuscles; a far greater structural resemblance

MECHANISTIC-MATERIALIST CONCEPTION

than had formerly been suspected is shown to exist between Light and Matter: and our conception of Nature has thus become the simpler, and also the loftier.

The nucleus of an atom having the atomic number N has, as we saw above, a positive charge equal to N times that of the proton, and in it practically the entire mass of the atom is concentrated. It had long been believed that the nuclei of atoms consist of protons and electrons, the number of protons exceeding that of the electrons by N, and practically the entire mass being due to the protons. This idea that the nucleus is of a complex nature was more or less enforced by the interpretation of radioactivity, the discovery of which was initiated by Henri Becquerel, and in essence was the work of Pierre Curie and of his wife and collaborator Marie Sklodovska, whose death was such a grievous blow to French Science. The radioactive substances are heavy elements, bearing the highest atomic numbers in the series of elements—from eighty-three to ninety-two. They are characterized by a spontaneous instability, that is, by the fact that from time to time the nucleus of one of their atoms explodes, at the same time changing into the nucleus of a lighter atom. This transformation is accompanied by the expulsion of electrons (β-rays), of the light atoms of Helium ($N=2$) (α-rays) and by extremely penetrative rays of very high frequency (γ-rays). For physicists the discovery of these radioactive phenomena was of extreme interest, since it proved to them that the nuclei are in fact complex structures, and that a complex nucleus in the process of disintegration gives rise to a simpler one—thus spontaneously realizing the transmutation of elements dreamed of by the alchemists of the Middle Ages. Unfortunately, however, radioactivity is a phenomenon on which we are unable to exert any influence, and which consequently we can merely observe without being able to modify the process. Some twenty years after the discovery of radio-activity a great step forward was taken, when Rutherford discovered artificial disintegration; for by bombarding light atoms with α-particles—which in turn are emitted by radioactive substances—it was proved possible to break up these light atoms; and in this way simpler atoms are obtained—a genuine artificial transmutation. The quantities of Matter which undergo this transmutation are naturally slight, yet it has at present substantial practical importance; theoretically, on the other hand, its interest is enormous, since it proves the unity of Matter and affords further knowledge on the structure of the nuclei.

This research into artificial transmutations has undergone considerable development in recent years, beginning in England, where, under the leadership of Rutherford, the physicists Chadwick, Cockcroft, Walton and Blackett have reached remarkable results, and later in the United States, where Lawrence's work may be mentioned. In France, Paris now possesses two very important centres where problems relating to nuclei are being pursued by young investigators of great ability. First we have the *Institut du Radium*, directed until her death by Madame Pierre Curie, and where Madame Joliot, *née* Curie, her husband Monsieur Joliot, Pierre Auger, Rosenblum and others are at work. And then there is the *Laboratoire de recherches physiques sur les rayons X*, founded and directed by the author's brother, where Jean Thibaud, J. J. Trillat, Leprince-Ringuet and others are, or were, pursuing skilled and fruitful studies.

I cannot here deal in any way with the details of the results obtained; these have led to a kind of nuclear chemistry, in which the transmutations are represented by means of equations strictly analogous to those long used by chemists to represent ordinary chemical reactions. I must, however, stress two fundamental discoveries made wholly unexpectedly in the course of these researches. The first of these is the discovery of the neutron; Chadwick and the Joliots independently discovered the presence, among the products of the process of disintegration, of a kind of corpuscle hitherto unknown. These corpuscles pass through Matter with great ease; they appear to have no electric charge, but to have a mass approximately equal to that of the proton. They are the neutrons, and there appears to be no doubt that they play an important part in the structure of the nuclei.

Within a year of the discovery of the neutron, in 1932, a fourth class of corpuscle was discovered in its turn. While studying the effects of the disintegration caused by cosmic rays, Anderson, and also Blackett and Occhialini, independently demonstrated the existence of positive electrons—*i.e.*, corpuscles having the same mass as the electron and with an electric charge equal to that of the electron, but bearing an opposite sign. These positive electrons, which are a great deal rarer than the negative, appear to play an important part in the phenomena connected with the nuclei.

The upshot of these recent sensational discoveries was to leave the position a good deal more complicated than it had ever been, since we now know four different kinds of corpuscles—electrons, protons, positive electrons and neutrons. The question one asks is whether they all are in fact elementary; and the answer is undoubtedly in the negative. It would appear that one of the four must be complex. It may be assumed, for example, that the proton, the electron and the positive electron are the elementary units, in which case the neutron consists of a proton to which is due almost the entire mass of the neutron, and of an electron which neutralizes the charge of the proton. Or again one may assume—and this appears to me the more attractive hypothesis—that it is the neutron and the two kinds of electron which are the elementary corpuscles, in which case the proton would consist of a neutron and a positive electron, and would cease to rank as a simple corpuscle. In any case the discovery of the neutron and of the positive electron are valuable additions to our knowledge of the atomic world.

A word may here be said about cosmic rays. A series of experiments undertaken during recent years, the most important of which are those carried out by Millikan, has proved the existence of extremely penetrative rays which appear to come from interplanetary space. It has been found, too, that these rays have extremely powerful effects on Matter and cause various kinds of atomic disintegration. Research into cosmic rays is difficult, and as yet little is known of their nature, but there is small doubt that numerous interesting results will shortly be obtained in this respect as well.

All too brief as this survey is, it will have shown that laboratory research during the last few years has led to results of the utmost interest almost each day. But theoretical Physics, too, whose function it is to provide a guiding light for experimental Physics, has not remained idle.

In the history of theoretical Physics, then, during the last thirty years, there are two great landmarks: the Theory of Relativity and the Quantum Theory, two doctrines of the widest scope; and while the Theory of Relativity is less closely connected with the advancement of atomic Physics, it is the more familiar to the man in the street. Its origin lies in certain phenomena of the propagation of Light which could not be explained by the older theories; but by an intellectual effort which will always hold an eminent place in the annals of Science, Einstein removed the difficulty by the introduction of entirely novel ideas on the nature of Space and Time and their interrelation. Hence the origin of that remarkable Theory of Relativity, which later achieved an even more general scope by providing us with an entirely new conception of Gravitation. It is true that certain of the experimental verifications of the Theory have been, and still remain, in debate; but it is quite certain that it provides us with extremely novel and fertile points of view. For it has shown how the removal of certain preconceived ideas, adopted through habit rather than logic, made it possible to overcome obstacles regarded as insuperable and thus to discover unexpected horizons; and for physicists the Theory of Relativity has been a marvellous exercise in overcoming mental rigidity.

The Quantum Theory and its developments, if less generally familiar, are certainly at least equally important, since by means of this Theory it has been possible to make use of the discoveries of experimental Physics to form a science of atomic phenomena. When a more precise description of these phenomena was felt to be necessary, the fundamental fact which became apparent was that it was imperative to introduce completely novel concepts which had been entirely unknown to classical Physics. For in order to describe the atomic world it is not enough to transport the methods and images which are valid on the human, or on the astronomical scale, to another and very much smaller scale. We saw that, following Bohr, scientists succeeded in imagining atoms to be miniature solar systems in which the electrons played the part of the planets, and in tracing their orbits round a central sun bearing a positive charge. But if this image was to give really valuable results, it became necessary to assume, still further, that the atomic solar system obeyed Quantum Laws; and these were entirely different from the Laws governing the systems with which Astronomy deals. The more carefully this difference was considered, again, the more its wide scope and fundamental significance began to be appreciated; for the intervention of quanta brought about the introduction of discontinuity in atomic Physics, and this introduction is of essential importance, since without it atoms would be unstable and Matter could not exist.

We saw that the discovery of the double nature of electrons, as at once corpuscular and undulatory, was followed by a change in the Quantum Theory, so that this was given a new form, some years ago, called Wave Mechanics. The new form has met with manifold success, and Wave Mechanics has brought about a better understanding and prediction of those phenomena which depend upon the existence of quantized stationary states for atoms. Every branch of Science, including Chemistry, has benefited from the impetus due to the new theory, because this has brought with it an entirely novel and interesting manner of interpreting chemical combinations.

The development of Wave Mechanics, then, has compelled physicists to give an ever wider and wider scope to their concepts. For according to the new principles, the Laws of Nature no longer have the strict character which they bear in classical Physics: phenomena (in other terms) are no longer subject to a rigorous Determinism; they only obey the Laws of Probability. The famous Principle of Uncertainty advanced by Heisenberg gives an exact formulation to this fact. Even the notions of Causality and of Individuality have had to undergo a fresh scrutiny, and it seems certain that this major crisis, affecting the guiding principles of our physical concepts, will be the source of philosophical consequences which cannot yet be clearly perceived.

> Louis de Broglie: *Matter and Light*, translated by W. H. Johnston. (Quoted by kind permission of Messrs. Allen & Unwin Ltd.)

Bibliography:

L'électron magnétique, Paris, 1934; *Ondes et corpuscles*, Paris, 1928; Matière et lumière, Paris, 1937 (*Matter and Light*, Allen & Unwin); Continu et discontinu du physique moderne, Paris, 1941; *Physique et Microphysique*, Paris, 1947 (*Physics and Microphysics*, Hutchinson); Théorie génerale des particules à spin, Paris, 1943; Editor of *La cybernétique, théorie du signal et de l'information*, Paris, 1951.

SUMMARY

In this section on Historical Sources I have tried to give at least an outline of the development of modern science by quoting a necessarily limited selection from the writings of the most prominent exponents. By way of summary I should like to stress the following:

1. Modern science, in its beginnings, was characterized by a conscious modesty; it made statements about strictly limited relations that *are only valid within the framework of these limitations.*
2. *This modesty was largely lost during the nineteenth century.* Physical knowledge was considered to make assertions about nature as a whole. Physics wished to turn philosopher, and the demand was voiced from many quarters that all true philosophers must be scientific.
3. Today physics is undergoing a basic change, the most characteristic trait of which is a return to its original self-limitation.
4. The philosophic content of a science is only preserved if science is conscious of its limits. Great discoveries of the properties of individual phenomena are possible only if the nature of the phenomena is not

generalized *a priori*. Only by leaving open the question of the ultimate essence of a body, of matter, of energy, *etc.*, can physics reach an understanding of the individual properties of the phenomena that we designate by these concepts, an understanding which alone may lead us to real philosophical insight.

ABOUT THE AUTHOR

Werner Carl Heisenberg was born in Würzburg on the 15th December, 1901, son of Dr. August Heisenberg, then teacher at the Gymnasium. In 1909 his father became Professor of Middle and Modern Greek at the University of Munich. Here Werner Heisenberg attended the Maximilian Gymnasium from which he matriculated in 1920. He then studied physics at Munich. His teachers were chiefly Sommerfeld, Wien, Pringsheim and Rosenthal. During the winter term of 1922-23 he studied under Born, Frank and Hilbert in Göttingen, and in 1923, working under Sommerfeld in Munich, he obtained his Doctorate of Philosophy. He then became an assistant of Born in Göttingen. In the summer of 1924 he gained the *venia legendi* at Göttingen. In the winter of 1924-25 he worked as a Rockefeller Scholar under Niels Bohr in Copenhagen. In the summer of 1925 he again worked in Göttingen. In 1926 he was appointed lecturer in theoretical physics at the University of Copenhagen. In the autumn of 1927 he became Professor-in-Ordinary of theoretical physics at the University of Leipzig. In 1929 he went on a long lecture tour of the U.S.A., Japan and India. In 1932 and 1939 he again lectured in the U.S.A. In 1933 he

BIBLIOGRAPHY

ABRO, A. D', *Revolution of Scientific Thought from Newton to Einstein*. (1950.)

BAVINK, B., *Die Hauptfragen der heutigen Naturphilosophie*. (1928.)
Ergebnisse und Probleme der Naturwissenschaften. (1953.)
Was ist Wahrheit in den Naturwissenschaften? (1947.)
Das Weltbild der heutigen Naturwissenschaften und seine Beziehung zu Philosophie und Religion. (1947.)
The Anatomy of Modern Science. (English translation, 1932.)
Science and God. (English translation, 1933.)

BECHER, E., *Naturphilosophie*. (1914.)

BOHR, N., *The Penetration of Atomic Particles Through Matter*. (1948.)

BOPP, F., and RIEDEL, O., *Die Physikalische Entwicklung der Quantentheorie*. (1950.)

BORN, M., *Natural Philosophy of Cause and Chance*. (1949.)
The Restless Universe. (1951.)

BOUTROUX, E., *De la contingence des lois de la nature*. (Acad. Fr.)

BRIDGMAN, P. W., *The Logic of Modern Physics*. (1927.)
The Nature of Physical Theory. (1956.)

BROGLIE, L. DE, *Matter and Light; The New Physics*. (English translation, 1939.)
The Revolution in Physics. (English translation, 1954.)
Physics and Microphysics. (English translation, 1955.)

BURKAMP, W., *Naturphilosophie der Gegenwart*. (1930.)

BURTT, E. C., *The Metaphysical Foundations of Modern Physical Science*. (1925.)

BUTTERFIELD, H., *The Origins of Modern Science*. (1949, 2nd ed., 1957.)

CASSIRER, E., *The Philosophy of Symbolic Forms*. Vol. 4: *The Problem of Knowledge*. (1950.)
The Philosophy of History. (1936.)

CONRAD-MARTIUS, H., *Der Selbstaufbau der Natur*. (1944.)

DAMPIER, W. C., *The History of Science*. (1942.)

DANNEMANN, F., *Aus der Werkstatt grosser Forscher*. (1922.)
Die Naturwissenschaften in ihrer Entwicklung. (1923.)
Vom Werden der Naturwissenschaftlichen Probleme. (1928.)

DESSAUER, F., *Mensch und Kosmos*. (1949.)
Die Teleologie und die Natur. (1949.)
Am Rande der Dinge. Über das Verhältnis von Wissen und Glauben. (2nd Ed., 1952.)
Religion im Lichte der heutigen Naturwissenschaft. (3rd ed., 1952.)

DINGLER, H. (Edited): *A Century of Science, 1851–1951*. (1951.)

DREYER, J. L., *History of the Planetary System from Thales to Kepler*.

DRIESCH, H., *Die Maschine und der Organismus*. (Bios., Vol. IV, 1935.)
The Science and Philosophy of the Organism. (English translation, 1900.)

DUHEM, P., *The Aim and Structure of Physical Theory*. (English translation, 1954.)

EDDINGTON, A. S., *The Nature of the Physical World*. (1935.)
New Pathways in Science. (1935.)
The Philosophy of Physical Science. (1939.)

EINSTEIN, A., *Geometrie und Erfahrung.* (1921.)
The Meaning of Relativity. (1951.)
EINSTEIN, A., and INFELD, L., *The Evolution of Physics.* (1938.)
FRANK, P., *Das Ende der Mechanistischen Physik.* (1935.)
GEIGER, M., *Die philosophische Bedeutung der Relativitätstheorie.* (1921.)
GRASSI, E., and UEXKULL, TH. VON, *Wirklichkeit als Geheimnis und Auftrag.* (1951.)
GROOT, H., *Kosmologische theorieën vorheen en thans.* (1947.)
HAAS, H., *Einführung in die theoretische Physik.* (1930.)
HARTMANN, H., *Schöpfer des neuen Weltbildes.* (1952.)
HARTMANN, M., *Die philosophischen Grundlagen der Naturwissenschaften.* (1948.)
HARTMANN, M., and GERLACH, W., *Naturwissenschaftliche Erkenntnis und ihre Methoden.* (1937.)
HARTMANN, N., *Der Aufbau der realen Welt.* (1940.)
HAZARD, P., *La crise de la conscience européenne; 1690–1715.* (1935.)
HEIM, K., *Der christliche Gottesglaube und die Naturwissenschaft.* (Vols. 1 and 2, 1949 and 1951.)
HEISENBERG, W., *Die physikalischen Prinzipien der Quantentheorie.* (4th ed., 1944.)
Die Physik der Atomkerne. (8 Lectures. 3rd ed., 1949.)
Die Einheit des Naturwissenschaftlichen Weltbildes, in *Wandlungen in den Grundlagen der Naturwissenschaft.* (8th ed., 1948.)
Die gegenwartige Aufgabe der theoretischen Physik. (Scientia, February, 1938.)
Die Rolle der Unbestimmtheitsrelation in der modernen Physik. (Monatshefte für Mathematik und Physik. Vol. 38, No. 2, 1931.)

HEISENBERG, W., *Prinzipielle Fragen der modernen Physik*, in *Wandlungen in den Grundlagen der Naturwissenschaft*. (8th ed., 1948.)
The Philosophical Principles of Quantum Theory. (English translation, 1930.)
The Present Situation in the Theory of Electric Particles. (English translation, 1949.)
Electron Theory of Superconductivity. (English translation, 1930.)
Philosophic Problems of Nuclear Science. (English translation, 1952.)
Nuclear Physics. (English translation, 1953.)

HELMHOLTZ, H. VON, *Schriften zur Erkenntnistheorie*. (1921.)

HILDESHEIMER, A., *Die Welt der ungewohnten Dimensionen*. (1953.)

HÖNIGSWALD, R., *Naturphilosophie*. (Jahrbücher der Philosophie, Year 1, 1913.)

JAKOB, H., *Die Grundlagen unserer naturwissenschaftlichen Erkenntnis*. (1948.)

JEANS, J., *The Universe Around Us*. (1931.)
Physics and Philosophy. (1942.)
The Growth of Physical Science. (1947, 2nd ed., 1951.)

JOAD, C. E. M., *Philosophical Aspects of Modern Science*. (3rd ed., 1943.)

JORDAN, P., *Die Physik und das Geheimnis des organischen Lebens*. (6th ed., 1948.)
Das Bild der modernen Physick. (1947.)
Schwerkraft und Weltall. (1952.)
Physics in the 20th century. (Translation.)

LAUE, M. VON, *History of Physics*. (English translation, 1950.)

LOTZE, R., and SIHLER, H., *Das Weltbild der Naturwissenschaft*. (1954.)

MACH, E., *Die Analyse der Empfindungen*. (1922.)

MARCH, A., *Natur und Erkenntnis*. (1948.)

MARGENAU, H., *The Nature of Physical Reality*. (1950.)

MIELI, A., *La science arabe et son role dans l'évolution scientifique mondiale*. (1939.)

NIGGLI, P., *Probleme der Naturwissenschaften*. (1949.)

OPPENHEIMER, J. R., *Science and the Common Understanding*. (1953.)

PLANCK, M., *Die Einheit des Wissenschaftsbildes*. (1909.)
Sinn und Grenzen der exakten Wissenschaft. (2nd ed., 1947.)
Vorträge und Erinnerungen. (5th ed., 1949.)
Kausalagesetz und Willensfreiheit. (1923.)
Das Weltbild der neuen Physik. (10th ed., 1947.)
Wege zur physikalischen Erkenntnis. (1944.)
Wissenschaft und Methode. (1912.)
The Universe in the Light of Modern Physics. (English translation, 1937.)
Scientific Autobiography and other Papers. (English translation, 1950.)

POINCARÉ, H., *La Valeur de la Science*. (1904.)
Science and Hypothesis. (English translation, 1905.)
Dernière Pensées. (1913.)

REICHENBACH, H., *Philosophic Foundations of Quantum Mechanics*. (English translation, 1944.)

SARTON, G., *Introductions to the History of Science*. (1947–1950.)

SELIGER, R., *Die Grundbeziehungen der neuen Physik*. (1948.)

TAYLOR, F. S., *A Short History of Science*. (1939.)
The World of Science. (2nd ed., 1950.)
An Illustrated History of Science. (1956.)

THOMSON, G., *The Atom*. (1947.)

WEIZSÄCKER, C. F. VON, *The World View of Physics.* (English translation, 1949.)
The History of Nature. (English translation, 1951.)

WEIZSÄCKER, C. F. VON, and JUILFS, J., *Contemporary Physics.* (English translation, 1957.)

WENZL, A., *Metaphysik der Physik von heute.* (1935.)

WHITEHEAD, A. N., *The Concept of Nature.* (1920.)
An Inquiry Concerning the Principle of Natural Knowledge. (1925.)
Science and the Modern World. (1927.)

NAME INDEX

ALEMBERT, Jean Lerond d', 129ff
Anderson, Carl David, 175
Archimedes, 86
Archytas, 91
Aristotle, 33
Auger, Pierre, 174

BACON, Francis, 110
Becquerel, Henri, 173
Blackett, Patrick Maynard Stuart, 174
Bohr, Niels, 16, 39, 40f, 163ff, 171, 177
Boltzmann, Ludwig, 37f
Born, Max, 182
Boyle, Robert, 37, 110, 126f
Broglie, Louis de, 152, 158ff
Broglie, Maurice de, 173

CHADWICK, Sir James, 174f
Chuang-Tzu, 20f
Cicero, 79
Cockcroft, Sir John Douglas, 174
Columbus, 65
Copernicus, 77, 83
Cotes, R., 111ff
Curie, Marie, 173
Curie, Pierre, 173
Cusanus, 79

DAVISSON, Clinton Joseph, 172
Democritus, 12, 35f, 61
Descartes, René, 29, 62, 124, 126f, 148
Diodati, Elia, 88
Du Bois-Reymond, Emil, 148

EDDINGTON, Sir Arthur Stanley, 153
Einstein, Albert, 39, 47, 176
Epicurus, 126
Euclid, 56f, 82f

FRANK, Ph., 182
Fresnel, Augustin, 167ff
Freyer, Hans, 66

GALEN, 97
Galilei Galileo, 8f, 85ff
Gassendi, Petrus, 61, 126f
Germer, Lester Halbert, 172
Gibbs, Josiah Willard, 37f
Gilbert, William, 110

HAHN, Otto, 45
Harvey, William, 110
Heisenberg, Werner, 178, 182f
Hertz, Heinrich, 152ff
Hilbert, David, 182
Hohenburg, H. von, 77
Huygens, Christian, 121ff, 135, 166

191

NAME INDEX

Joliot, F., 174f
Joliot-Curie, Irene, 174f
Jordan, Pascual, 42

Kamlah, Wilhelm, 9
Kant, Immanuel, 33f, 146
Kepler, Johannes, 8, 71ff, 85, 105

Lagrange, Joseph Louis, 157
Laplace, Pierre Simon de, 34
Lawrence, Ernest Orlando, 174
Leibnitz, Gottfried Wilhelm, 147f
Leonardo da Vinci, 87
Leprince-Ringuet, 174
Leucippus, 35, 61
Liebig, Justus von, 137f
Locke, John, 128
Lucretius, 76

Mariotte, Edme, 110
Mendelejeff, Dimitri Ivan, 164
Mettrie, Julian Offray de la, 135ff
Millikan, Robert Andrews, 176

Newton, Sir Isaac, 9ff, 34, 57, 110ff

Occhialini, 175
Ostwald, Wilhelm, 137ff

Perrin, Jean, 161
Planck, Max, 38f, 63, 164f
Plato, 59f, 79, 83
Ponte, 172
Pringsheim, Ernst, 182
Pythagoras, 56f, 73, 77

Rhaeticus, 83
Rosenblum, 174
Rosenthal, 182
Rutherford, Lord Ernest of Nelson, 163, 174

Schelling, Friedrich Wilhelm, 137
Socrates, 90ff
Sommerfeld, Arnold, 39, 59, 182
Spinoza, Baruch, 147

Thibaud, Jean, 174
Thomson, George P., 172
Torricelli, Evangelista, 110
Trillat, J. J., 174

Voltaire, 128

Walton, Ernest Thomas Sinton, 174
Wien, Wilhelm, 182

Young, Thomas, 166, 167

OHIO UNIVERSITY LIBRARY

Please return
have fini
fine i